ANALYTICAL METHODS IN PETROLEUM UPSTREAM APPLICATIONS

ANALYTICAL METHODS IN PETROLEUM UPSTREAM APPLICATIONS

edited by
César Ovalles
Carl E. Rechsteiner Jr.

CRC Press
Taylor & Francis Group
Boca Raton London New York

CRC Press is an imprint of the
Taylor & Francis Group, an **informa** business

CRC Press
Taylor & Francis Group
6000 Broken Sound Parkway NW, Suite 300
Boca Raton, FL 33487-2742

First issued in paperback 2020

© 2015 by Taylor & Francis Group, LLC
CRC Press is an imprint of Taylor & Francis Group, an Informa business

No claim to original U.S. Government works

ISBN 13: 978-0-367-57594-6 (pbk)
ISBN 13: 978-1-4822-3086-4 (hbk)

Library of Congress Cataloging-in-Publication Data

Analytical methods in petroleum upstream applications / edited by Cesar Ovalles and
    Carl E. Rechsteiner, Jr.
        pages cm
    "A CRC title."
    Includes bibliographical references and index.
    ISBN 978-1-4822-3086-4 (hardcover : alk. paper) 1. Petroleum--Refining. 2.
    Petroleum--Analysis. I. Ovalles, C?sar. II. Rechsteiner, Carl E., Jr.

    TP690.A736 2015
    665.5--dc23

                                                                              2014043097

Visit the Taylor & Francis Web site at
http://www.taylorandfrancis.com

and the CRC Press Web site at
http://www.crcpress.com

Contents

SECTION I Background Chapters

SECTION II Water Analysis

SECTION III Properties

SECTION IV   Analytical Measurements

SECTION V   Heavy Ends and Asphaltenes

SECTION VI   Modeling and Chemometrics

Foreword

Analytical Methods in Petroleum Upstream Applications crosses the boundary from exploration- and production-related chemistry to refining, from upstream through downstream. For decades, in upstream, oil has been treated only as a fluid with flow properties and pressure–volume–temperature (PVT) responses—how do we get this product to move through the reservoir and to the surface tanks? Once in the pipeline, "it is someone else's problem." Production engineers inherited from earth scientists whatever the crude was, and field hands got it to the surface any way they could. Reservoir engineers reduced oil to viscosity and pour point to run their models. Chemical engineers had to resolve any well and surface problems with little or no data on oil/water properties. Then upstream delivered their "oil" to downstream. It worked, but not optimally. Value has been lost along the value chain.

Advances in laboratory and field measurements, on-line and in-line, enable a team to optimize the production and valuation of oil from early field development through enhanced oil recovery, through design of surface facilities and pipelines, to shipment of oil to specific refineries for highest return on investment. In this book, the entire hydrocarbon value chain from the prospect phase to the refinery and sale points has been linked through chemistry with well-defined analytical methods. This book gives the reader full coverage of the value chain enhancement provided by geochemistry, modern analytical chemistry, and data processing methods.

Dr. M. Boduszynski's seminal work on continuity, definition and modeling of crudes, and discussions on specific analytical procedures for the field and laboratory, which are coupled with water chemistry, scale/corrosion, and flow issues, to help the dedicated geological and engineering team to define product value in the ground and optimize production through facilities. In the oil business, we often "correlate" and estimate properties to try to determine oil properties in the reservoir and facilities. Nuclear magnetic resonance actually measures oil properties whether with logging tools in the reservoir or measurements in the laboratory. Measured viscosities and other flow properties can then be entered into reservoir simulation. In addition, critical components that inhibit production, such as asphaltenes, can be measured, and systems installed that are designed to mitigate production and pipeline problems.

This book carries the analytical information to the final steps of chemometrics. This process can be used to optimize the value chain sectors from production, through transportation, to refining. This book will enable the inquiring earth scientist or engineer to develop a basic understanding of critical chemical issues that can significantly influence the delivery of top value for the crudes they discover, produce, transport, and refine.

Dr. Paul Henshaw

Preface

With the rapid advances in measurement technologies in recent years, the editors decided to assemble a book detailing such advances that are germane to the upstream portion of the petroleum industry. The breadth of application of these technologies within the petroleum industry means that we will include some non-upstream applications as appropriate.

Chapter 1 of this book, by Dr. Mieczyslaw M. Boduszynski, sets the stage for the remaining chapters by discussing the composition of petroleum through the use of his "petroleum molecular composition continuity model." This model provides a context for the analytical measurements discussed later.

To obtain accurate measurements for any complex material, it is paramount that the sample is collected in as close to native conditions as practical. Chapter 2, by William M. Cost, describes a modern modular sampling system that can be used in either the laboratory or in the process area to collect and control samples for subsequent analysis. Such systems can also be used with process analytical sensors, etc. to provide information on flowing streams.

Darrell L. Gallup provides Chapter 3 on oil-in-water monitoring. Practitioners unfamiliar with petroleum operations may not realize the importance of oil-in-water measurements. In most upstream production facilities, water is produced at a rate that is almost an order of magnitude greater than the accompanying oil. Effective separation of the oil from the water can have significant financial impact, both for producing salable oil as well as mitigating environmental impacts.

Chapters 4 and 5 deal with the chemical and physical properties of heavy oils, their fractions, and products from their upgrading. Chapter 4 by Francisco Lopez-Linares et al. compares and contrasts the characterization of two vacuum residues from a Canadian and a Saudi Arabian crude and the products of subjecting those residua to thermal cracking (visbreaking) conditions. Chapter 5 by Lante Carbognani Ortega et al. deals specifically with the analysis of olefins (unsaturated, nonaromatic hydrocarbons) in heavy oils, bitumen, and their upgraded products. Olefin characterization is of paramount importance because of their negative processing impacts leading to stability issues, such as polymerization and deposit formation.

Chapters 6 through 8 deal with workhorse analytical measurements. Chapter 6 by Carl E. Rechsteiner Jr. et al. discusses advances in gas chromatography (GC). GC continues to be a workhorse for petroleum analyses. This chapter provides a review of the GC application space and introduces a compact, near-research-grade GC that is capable of a wide range of analytical measurements. This system has a throughput of approximately 10 times greater than conventional GCs, with much lower power consumption and reduced cooling requirements.

Chapter 7 by Zheng Yang et al. chronicles the development of nuclear magnetic resonance (NMR) applications from simple well logging to measuring critical reservoir parameters. The properties studied for hydrocarbons and rock formations include mineralogy-independent porosities, irreducible water saturations, permeabilities, and viscosities.

Chapter 8, by Teresa E. Lehmann and Vladimir Alvarado, also discusses NMR technology from a different perspective than the prior chapter. Much of this chapter deals with the analysis of the collected data to improve understanding of how to produce oil- and gas-containing reservoirs.

Chapters 9 through 11 focus on asphaltene and heavy ends analysis. Chapter 9 by Estrella Rogel et al. discusses several new methods to analyze these materials by on-column filtration. These methods, on-column filtration, asphaltene solubility profile, and separation of asphaltenes in solubility fractions, have been applied to a number of crudes and derived fractions to rapidly determine asphaltene content, measure the stability of heavy oil/naphtha blends and crude oils during steam and CO_2 flooding, and monitor changes in asphaltene behavior during production activities.

Chapter 10, by Estrella Rogel and Michael Roye, describes the absorption of asphaltenes on iron oxide. In particular, this chapter discusses measurements that allow determining isotherms from various solvents, pentane to toluene, at several temperatures, and the kinetics of their absorption onto iron oxide.

Chapter 11, by Farshid Mostowfi and Vincent Sieben, demonstrates the measurement of asphaltenes with microfluidic technology. The reduced size and the resulting reduction in solvent usage make this technology an intriguing approach for future use.

The final three chapters apply chemometrics and modeling approaches to improve the understanding of upstream (and mid- and downstream) operations. Chapter 12 by Thomas I. Dearing et al. shows the value of data fusion to feed chemometric analysis. Inferring physical and chemical properties of petroleum products from spectroscopic measurements is well known; for example, the prediction of octane numbers, aromatics content, etc., in motor gasoline by infrared spectroscopy is widely used in refining operations. This chapter demonstrates an approach where multiple spectroscopic measurements (from instruments whose fundamental measurement basis are different, i.e., orthogonal data sets in the mathematical sense) are obtained on the same materials and then the data are fused to form a single, unbiased data set for chemometric analysis. Predictions made with this approach are substantially better than those obtained from a single spectroscopy.

Chapter 13 by Jan-Willem Handgraaf et al. describes the use of computer simulations to compute the properties of oil–brine–surfactant microemulsions. The importance of this approach is to understand the effectiveness of different surfactants for promoting enhanced oil recovery.

Chapter 14 by Jorge A. Orrego-Ruiz et al. describes the use of chemometrics (principal component analysis) on infrared spectroscopic data to develop molecular information for a set of crude oils from Colombia. Improved understanding of the molecular nature of these crudes and their fractions allows optimization of subsequent processing for refining.

The editors would like to thank all of the contributing authors for their efforts and hope that this is useful to you, the reader.

César Ovalles
Chevron Energy Technology Company
COvalles@Chevron.com

Carl E. Rechsteiner Jr.
CRechsteiner Consulting LLC
Carl.Rechsteiner@gmail.com

For any questions concerning the content of this material, please contact the editors at the above listed e-mails.

MATLAB® is a registered trademark of The MathWorks, Inc. For product information, please contact:

The MathWorks, Inc.
3 Apple Hill Drive
Natick, MA 01760-2098 USA
Tel: 508 647 7000
Fax: 508-647-7001
E-mail: info@mathworks.com
Web: www.mathworks.com

Acknowledgments

We thank Chevron Energy Technology Company for providing permission to publish this book and some of the individual chapters.

The editors also thank all of the contributing authors for their efforts and dedication to making this project a successful endeavor.

Editors

César Ovalles is technical team leader at Chevron Energy Technology Company in Richmond, California. He earned a BSc in chemistry at Simon Bolivar University and a PhD in the same field at Texas A&M University. He worked for 16 years at Petróleos de Venezuela Sociedad Anónima–Instituto de Tecnología Venezolano del Petróleo (PDVSA–INTEVEP). In 2006, he joined Chevron to work in research and development (R&D) in petroleum chemistry and characterization of heavy and extra-heavy crude oils and their fractions. He is also involved in new methods of asphaltene analysis and in R&D in the chemistry of heavy and extra-heavy crude oil upgrading processes. He currently supervises a nine-member team in a variety of projects.

During 25 years of industrial experience, César has published 34 papers in peer-reviewed scientific journals, he has been awarded 16 patents, and he has presented 81 papers at scientific and technical conferences. Additionally, he has published 14 articles in Venezuelan journals and 85 technical reports for a total of 230 total scientific productions. César also served as associate editor of *Revista de la Sociedad Venezolana de Catalisis* from 1996 to 2000 and *Vision Tecnologica* (technical journal of PDVSA–INTEVEP) from 2000 to 2002. He currently serves as referee of scientific articles for several Venezuelan (*Interciencia, Acta Científica*, etc.) and international (*Energy & Fuels*, Fuels, *SPE-Reservoir Engineering, Journal of Canadian Petroleum Technology*, etc.) journals. César is married to his college sweetheart, Luisa Elena, and is the father of two grown children. His son, César Arturo, is a computer science graduate working for Safeway Supermarkets, and his daughter, Manuela, is a microbiology major at San Francisco State University.

Carl E. Rechsteiner Jr. is the owner of CRechsteiner Consulting LLC in Petaluma, California, providing petroleum composition and measurement advice to a number of companies, including both established oil companies and new, start-up instrument vendors. He earned BS degrees in applied mathematics and chemistry at California State Polytechnic University in San Luis Obispo and a PhD in analytical chemistry at the University of North Carolina at Chapel Hill. His industrial career began at Arthur D. Little Inc. in Cambridge, Massachusetts, where he spent four years dealing with environmental, food, and flavor measurement issues, interacting with regulatory agencies and participating in or leading studies involving mergers and acquisitions. In 1981, Dr. Rechsteiner joined Chevron Research Company in Richmond, California, where he spent 31 years in a number of roles involving numerous measurement technologies for elucidating petroleum compositions. He managed a number of research and development projects that developed and implemented measurement technologies (especially in the chromatographic and spectroscopic sciences) across Chevron's Global Downstream laboratory organization, and created an infrastructure to support data-rich process analyzers within Chevron's operations.

Dr. Rechsteiner has 24 refereed publications, has coauthored six books dealing with measurement of organic compounds, and has made more than 60 presentations at meetings with the American Chemical Society, the International Forum on Process Analytical Chemistry, the Gulf Coast Conference, the Federation of Analytical Chemistry and Spectroscopy (FACSS), the International Society of Automation (ISA)—Analysis Division, the First International Microtechnology Conference, the Pittsburgh Conference on Analytical Chemistry and Applied Spectroscopy, and the American Society for Mass Spectrometry.

Contributors

Vladimir Alvarado
Department of Chemical and Petroleum
　Engineering
University of Wyoming
Laramie, Wyoming

Mieczyslaw M. Boduszynski
Walnut Creek, California

Marten Buijse
Shell International Exploration and
　Production
Rijswijk, The Netherlands

Josune Carbognani
Chemical and Petroleum Engineering
　Department
Schulich School of Engineering
University of Calgary
Calgary, Alberta, Canada

William M. Cost
Parker Hannifin
Huntsville, Alabama

John Crandall
Falcon Analytical
Lewisburg, West Virginia

Thomas I. Dearing
Applied Physics Laboratory
University of Washington
Seattle, Washington

Mazin M. Fathi
Research and Development Center
Saudi Arumco
Saudi Arabia

Johannes G.E.M. Fraaije
Leiden Institute of Chemistry
Leiden University
and
Culgi BV
Leiden, The Netherlands

Darrell L. Gallup
Process Chemistry
Thermochem Inc.
Santa Rosa, California

Alexander Guzmán
Instituto Colombiano del Petróleo
Ecopetrol S.A.
Piedecuesta, Colombia

Jan-Willem Handgraaf
Culgi BV
Leiden, The Netherlands

Azfar Hassan
Schulich School of Engineering
University of Calgary
Calgary, Alberta, Canada

Shekhar Jain
Shell Technology Centre Bangalore
Bangalore, India

Teresa E. Lehmann
Department of Chemistry
University of Wyoming
Laramie, Wyoming

Francisco Lopez-Linares
Petroleum and Material
　Characterization Unit
Chevron Energy Technology Company
Richmond, California

Brian J. Marquardt
Applied Physics Laboratory
University of Washington
Seattle, Washington

Enrique Mejía-Ospino
Laboratorio de Espectroscopía Atómica
    y Molecular (LEAM)
Escuela de Química
Universidad Industrial de Santander
Bucaramanga, Colombia

Rachel Mohler
Chevron Energy Technology Company
Richmond, California

Michael E. Moir
Petroleum and Materials
    Characterization Unit
Chevron Energy Technology Company
Richmond, California

Daniel Molina
Laboratorio de Espectroscopía Atómica
    y Molecular (LEAM)
Escuela de Química
Universidad Industrial de Santander
Bucaramanga, Colombia

Farshid Mostowfi
Schlumberger DBR Technology Center
Edmonton, Alberta, Canada

Jorge A. Orrego-Ruiz
Laboratorio de Espectroscopía Atómica
    y Molecular (LEAM)
Escuela de Química
Universidad Industrial de Santander
Bucaramanga, Colombia

and

Instituto Colombiano del Petróleo
Ecopetrol S.A.
Piedecuesta, Colombia

Lante Carbognani Ortega
Schulich School of Engineering
University of Calgary
Calgary, Alberta, Canada

Pedro Pereira-Almao
Schulich School of Engineering
University of Calgary
Calgary, Alberta, Canada

Ajit Pradhan
Petroleum and Materials
    Characterization Unit
Chevron Energy Technology Company
Richmond, California

Estrella Rogel
Petroleum and Materials
    Characterization Unit
Chevron Energy Technology Company
Richmond, California

Ned Roques
Falcon Analytical
Lewisburg, West Virginia

Michael Roye
Chevron Oronite Company LLC
Richmond, California

Vincent Sieben
Schlumberger DBR Technology Center
Edmonton, Alberta, Canada

Boqin Sun
Petroleum and Materials
    Characterization Unit
Chevron Energy Technology Company
Richmond, California

Kunj Tandon
Shell Technology Centre Bangalore
Bangalore, India

Marianna Trujillo
Chemistry Department
Science Faculty
University of Calgary
Calgary, Alberta, Canada

Qiao Wu
Chemistry Department
Science Faculty
University of Calgary
Calgary, Alberta, Canada

Zheng Yang
Petroleum and Materials
    Characterization Unit
Chevron Energy Technology Company
Richmond, California

John Zintsmaster
Petroleum and Materials
    Characterization Unit
Chevron Energy Technology Company
Richmond, California

Section I

Background Chapters

1 Petroleum Molecular Composition Continuity Model

Mieczyslaw M. Boduszynski

CONTENTS

CONTEXT

Understanding petroleum composition provides the context for subsequent analytical measurements. This modern theory

- Describes petroleum composition on the basis of continuous changes as a function of the atmospheric equivalent boiling point (AEBP)
- Provides an organizing principle for relating molecular structure/properties to AEBP
- Defines the AEBP scale that encompasses the entire crude oil

ABSTRACT

The model is based on the concept of the continuous variation of the chemical composition and properties of petroleum as a function of the atmospheric equivalent boiling point (AEBP). The model provides an organizing principle that relates molecular structures and molecular weights to the AEBP. The hypothetical AEBP scale encompasses the entire crude oil, including the "nondistillable" residuum, allowing the comparison of crude oils and their fractions on a common, rational basis.

1.1 INTRODUCTION

A better understanding of the molecular composition of crude oil, and linking that knowledge to its physical and chemical behavior, is the key to better crude oil value assessments and predictions of the outcome of upstream and downstream operations.

This chapter discusses key aspects of the *petroleum molecular composition continuity model* proposed a number of years ago in a series of articles and in the book *Composition and Analysis of Heavy Petroleum Fractions* [1–6].

The model is based on the concept of the continuous variation of the chemical composition and properties of petroleum as a function of the atmospheric equivalent boiling point (AEBP). This concept can be applied in a wide range of subjects, from basic thermodynamics to phase behavior, process design, and conversion unit monitoring. It was influential in supporting the development of numerical methods used in engineering process simulators to routinely represent crude oil as a continuous mixture.

The model provides an organizing principle that relates molecular structures and molecular weights to the AEBP. The hypothetical AEBP scale encompasses the entire crude oil, including the "nondistillable" residuum, allowing the comparison of crude oils and their fractions on a common, rational basis. The continuity of changing molecular composition is important when interpolating and extrapolating the physical and chemical properties of crude oil fractions.

The most contentious model prediction states that heavy crude oil components, including asphaltenes, have a molecular weight of <2000 Da. It recognizes the existence of what we now call "archipelago"- and "island"-type molecules. All model predictions have been recently confirmed for the full range of crude oil components [7–11].

1.2 CRUDE OIL COMPOSITION CONUNDRUM

The unraveling of crude oil composition is a formidable challenge since there are hundreds of different crude oils in the world and no two crude oils are alike. The issue is further complicated by the fact that the traditional upstream and downstream approaches to crude oil composition differ, as shown in Figure 1.1.

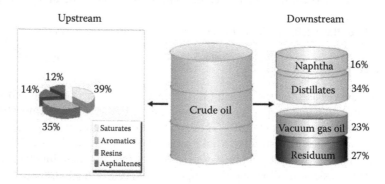

FIGURE 1.1 Upstream and downstream approaches to crude oil composition.

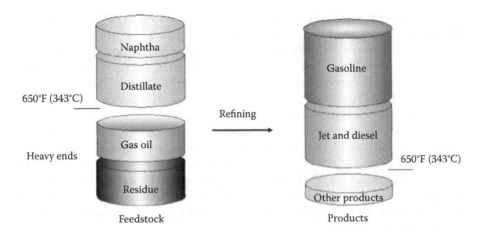

FIGURE 1.2   Crude oil as refinery feedstock.

The *upstream* approach defines the gross composition of a crude oil according to the content of the four group-type components, namely saturates, aromatics, resins, and asphaltenes (SARA). The first step in the so-called SARA analysis involves precipitation of asphaltenes (insolubles), which are operationally defined. Their content and composition depend on the solvent used (e.g., n-C5, n-C6, or n-C7) and the conditions of the precipitation procedure. The second step of SARA analysis involves the use of liquid chromatography to separate maltenes (solubles) into fractions of saturates, aromatics, and resins (also known as polars). Different chromatographic methods produce different results. The use of similar terms to describe the group-type components produced by different methods introduces further ambiguity.

The *downstream* approach defines the gross composition of crude oil in terms of the attainable yield and quality of the fractions produced by crude oil distillation as the primary refinery process. Further downstream processing involves conversion of feedstock molecules into new ones to meet the requirements of specific products.

Light transportation fuels are the highest-value petroleum products. Motor gasoline and jet and diesel fuels together account for approximately 90% of the crude oil consumption.

This is illustrated in Figure 1.2. Major conversion processes include catalytic cracking, hydrocracking, and coking. The difficulty of downstream processing determines the value of crude oil and requires molecular-level compositional information.

1.3 "LIGHT" AND "HEAVY" CRUDE OILS

Density is one of the most important properties of crude oils and their fractions, and it is a good indicator of crude oil quality. Density can be easily and precisely measured at a standard temperature, usually 60°F (15°C). Specific gravity is the ratio of the density of sample to the density of water at the same temperature.

The American Petroleum Institute (API) introduced the API gravity scale to expand the narrow range in specific gravity values. The API gravity is a modified inverse specific gravity with values ranging from about 50 for very "light" crudes to about ≤10 for very "heavy" crudes (see Figure 1.3). Thus, the term "heavy" means dense. Crude oils are ranked by the API gravity. Crude oil viscosity increases rapidly with decreasing API gravity (increasing density). The relation between crude oil API gravity and viscosity is illustrated in Figure 1.4.

Distillation is the primary refinery process that separates crude oil into progressively higher boiling fractions or "cuts" for further downstream processing. The API gravity values for crude oil fractions decrease with increasing boiling point. In other words, the deeper we cut, the heavier it gets. This is illustrated in Figure 1.5. The API gravities for the example crude oil range from a high value of about 70°API for the first naphtha cut to a negative value of −1.6°API for deep-cut vacuum residuum (nondistillable residuum).

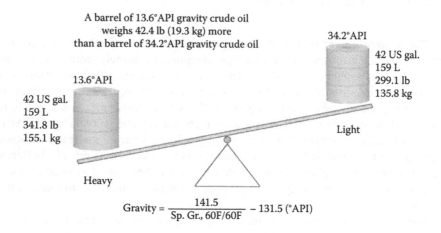

FIGURE 1.3   Crude oil API gravity.

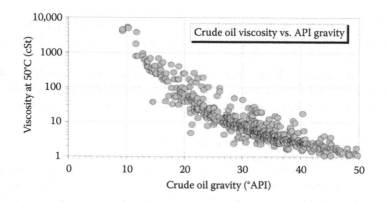

FIGURE 1.4   Crude oil API gravity and viscosity relation.

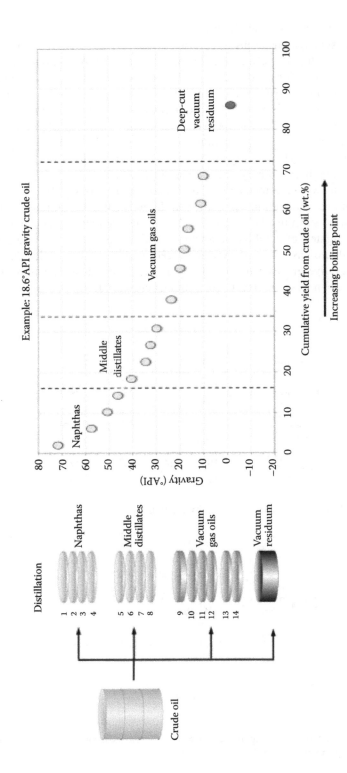

FIGURE 1.5 Effect of distillation on API gravity of crude oil fractions.

The adjectives "heavy" and "high molecular weight" are commonly but incorrectly used as synonymous terms to describe crude oils and their fractions. Figure 1.6 illustrates the dramatic effect of molecular structure on API gravity. The three compound types (paraffins, naphthenes, and aromatics) have the same carbon number but differ greatly in hydrogen content. The hydrogen-rich paraffin ($C_{18}H_{38}$) with the highest molecular weight of 254 is the lightest of the three compounds, having a gravity value of 48.6°API. The four-ring naphthene ($C_{18}H_{30}$), with the molecular weight of 242, has a lower gravity value of 12.0°API. Finally, the hydrogen-poor four-ring aromatic ($C_{18}H_{12}$), having the lowest molecular weight of 228, is the heaviest of the three, with a negative gravity value of −10.7°API.

Figure 1.7 shows three members of the *homologous series* C_nH_{2n-24}, spanning the carbon number range from C_{18} to C_{30} and the molecular weight range from 228 to 396. The alkyl-substituted chrysene $C_{30}H_{36}$, having the highest molecular weight of 396, is the lightest of the three compounds, with the gravity value of 6.3°API as compared with the unsubstituted chrysene $C_{18}H_{12}$ with the lowest molecular weight of 228 and negative gravity value of −10.7°API. The data further demonstrate that "heavy" does not necessarily mean "high molecular weight."

Structure	Formula	Z	API Gr.	Sp. Gr.
CH3−(CH2)16−CH3	$C_{18}H_{38}$ C_nH_{2n+2}	2	48.6	0.7857
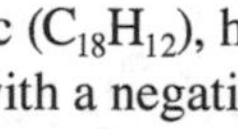	$C_{18}H_{30}$ C_nH_{2n-6}	−6	12.0	0.9861
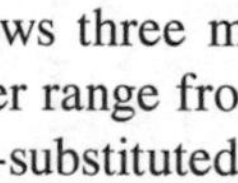	$C_{18}H_{12}$ C_nH_{2n-24}	−24	−10.7	1.1714

FIGURE 1.6   Molecular structure and gravity relation.

Structure	Formula	MW	API Gr.	Sp. Gr.
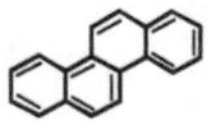	$C_{18}H_{12}$ C_nH_{2n-24}	228	−10.7	1.1714
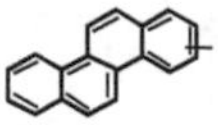 CH3	$C_{19}H_{14}$ C_nH_{2n-24}	242	−8.5	1.1504
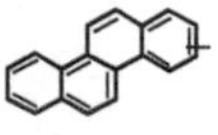 (CH2)11−CH3	$C_{30}H_{36}$ C_nH_{2n-24}	396	6.3	0.9574

FIGURE 1.7   Effect of increasing molecular weight on gravity for members of the "homologous series" of compounds with the general formula C_nH_{2n-24}.

1.4 DEEP-CUT DISTILLATION AND SEQUENTIAL EXTRACTION FRACTIONATION

A combination of the deep-cut distillation and sequential extraction (elution) fractionation (SEF) was used to separate the entire heavy crude oil into a series of "narrow" fractions down to the bottom of the barrel. This "volatility and solubility" separation scheme was developed to demonstrate the *continuity* of changing crude oil composition and properties.

The distillation scheme involved three distillation steps: (i) atmospheric distillation, (ii) vacuum distillation, and (iii) short-path distillation, also known as molecular distillation. This is illustrated in Figure 1.8.

Further separation of the nondistillable residuum into a series of solubility fractions was accomplished by using the SEF method illustrated in Figure 1.9. Detailed information on these separation methods can be found in earlier publications [1,2].

The underlying principles of both methods, distillation and SEF, are the various molecular interactions, called van der Waals forces. These consist of the dispersion (London) forces, permanent dipole–dipole forces, and hydrogen-bonding forces. For each homologous series of compounds, or for each compound type, the dispersion forces increase with molecular weight. As the dispersion forces increase, so does the boiling point, and so does the solubility parameter that governs the separation in liquid fractionations such as the SEF. Thus, the SEF can be considered as the "equivalent distillation."

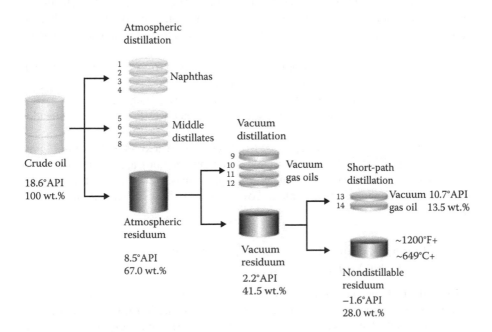

FIGURE 1.8 Deep-cut distillation.

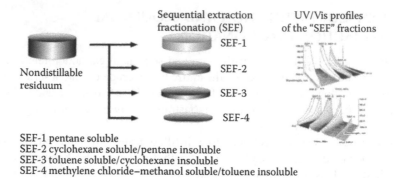

SEF-1 pentane soluble
SEF-2 cyclohexane soluble/pentane insoluble
SEF-3 toluene soluble/cyclohexane insoluble
SEF-4 methylene chloride–methanol soluble/toluene insoluble

FIGURE 1.9 Sequential extraction fractionation (SEF). (Reprinted with permission from *Energy & Fuels*, vol. 2, p. 597. Copyright 1988 American Chemical Society.)

1.5 HYPOTHETICAL AEBP SCALE

The development of the hypothetical AEBP scale, extending into the nondistillable residuum fractions, was necessary to demonstrate the continuity of changing crude oil composition and properties as a function of AEBP. The AEBP scale encompasses the entire boiling range of crude oil, starting with that of atmospheric pressure, continuing with those accessible by reduced pressure, and including the equivalent boiling ranges of nondistillable residuum solubility fractions. By this stratagem of including the nondistillable residue, an entire crude oil can now be described in terms of its various physical and chemical properties as they change with increasing AEBP.

The AEBP values for distillable fractions were derived from simulated distillation measurements and represent mid-boiling point values (at 50%). The AEBP values for the nondistillable residuum solubility (SEF) fractions were calculated using earlier developed correlations, involving vapor phase osmometry (VPO) average molecular weights [4,5].

The AEBP distribution curve for the example heavy crude oil is shown in Figure 1.10. The AEBP values for nondistillable residuum fractions are inflated owing to erroneously high VPO average molecular weight values. Nevertheless, these results allow for extending the AEBP scale to cover the entire heavy crude oil on a consistent basis.

1.6 CRUDE OIL ELEMENTAL COMPOSITION
AS A FUNCTION OF AEBP

Carbon and hydrogen are the major building blocks of petroleum molecules. Changes of carbon and hydrogen contents with increasing AEBP are illustrated in Figure 1.11. Contents of both elements decrease with increasing AEBP at the expense of increasing hetero-element content.

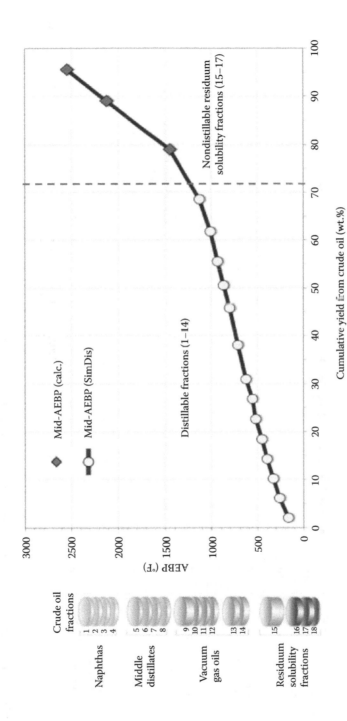

FIGURE 1.10 Example heavy crude oil AEBP distribution curve.

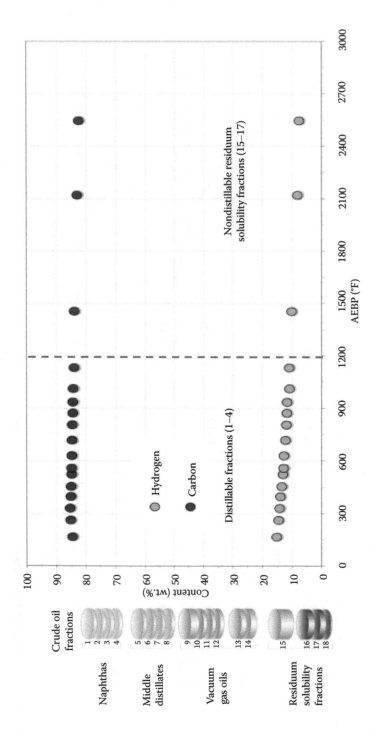

FIGURE 1.11 Carbon and hydrogen content variations as a function of AEBP.

Sulfur is the most abundant hetero-element in crude oils. The sulfur content in different crude oils varies, as shown in Figure 1.12. Heavy crudes have high sulfur content. The high cost of sulfur removal makes sulfur content the second most significant variable (after gravity) in the cost and value of crude oil.

Figure 1.13 compares changes of hydrogen and sulfur contents with increasing AEBP. The hydrogen and sulfur concentrations in the example whole crude oil are 11.5 and 3.3 wt.%, respectively. The hydrogen content decreases from a high value of about 15 wt.% for a low-boiling naphtha fraction to a low value of about 7 wt.% for the nondistillable residuum solubility fraction SEF-3 (#17). The hydrogen distribution pattern is consistent with the earlier-discussed API gravity changes (see Figure 1.5).

The sulfur content increases from a fraction of 1 wt.% in low-boiling distillate fractions to >7 wt.% in the nondistillable residuum solubility fraction SEF-3 (#17). The sulfur concentration follows a general sulfur distribution pattern in crude oils, where distillate fractions boiling up to 650°F (343°C) account typically for about 10% of the total sulfur content in crude oil. The remaining 90% of the total sulfur content in crude oil is usually evenly divided between the high-boiling (650–1200°F, 343–649°C) vacuum gas oil fractions and the nondistillable residuum (1200°F+, 649°C+).

The nitrogen content in crude oils is significantly lower than sulfur, ranging from a few parts per million (ppm) by weight to <1 wt.%. Figure 1.14 illustrates the nitrogen content in different crude oils. Heavy crude oils have a high nitrogen content. Changes in nitrogen content with increasing AEBP for a sample heavy crude oil are shown in Figure 1.15. The nitrogen content in the example whole crude oil is 0.37 wt.%. Low-boiling naphtha fractions are essentially free of nitrogen. Nitrogen concentration gradually increases from a few ppm in middle distillates to about 1.4 wt.% in the SEF-3 (#17) solubility fraction derived from the nondistillable residuum.

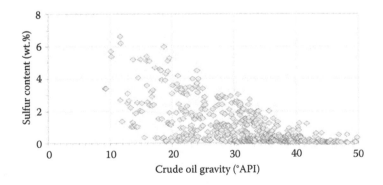

FIGURE 1.12 Sulfur content in different crude oils.

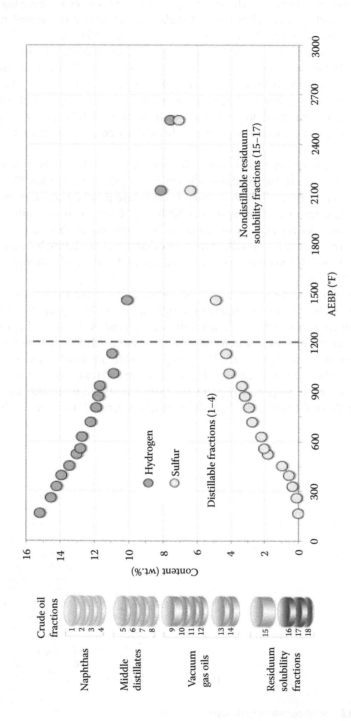

FIGURE 1.13 Hydrogen and sulfur contents in heavy crude oil fractions.

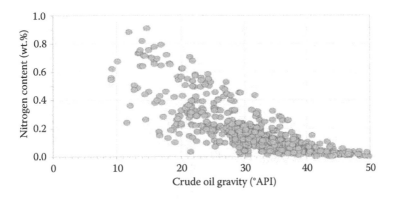

FIGURE 1.14   Nitrogen content in different crude oils.

Figure 1.16 provides a close look at nitrogen distribution in the middle distillate boiling range. The plot shows carbon and nitrogen chromatograms ("signatures") obtained by using gas chromatography simulated distillation (GC-SimDis) with nitrogen chemiluminescence detection.

Crude oil fractions boiling below 650°F (343°C) typically account for <1% of the total nitrogen content in crude oil. Nitrogen content in the vacuum gas oil boiling range (650–1200°F, 343–649°C) usually accounts for about 25% to 40% of the total nitrogen content in crude oil. The nondistillable residuum (1200°F+, 649°C+) contains the bulk of nitrogen, accounting for 60% to 75% of the total nitrogen content in crude oil.

Vanadium and nickel contents in different crude oils are shown in Figures 1.17 and 1.18, respectively. Most crude oils have a much higher vanadium than nickel content.

Variations of vanadium and nickel contents with the increasing AEBP are illustrated in Figures 1.19 and 1.20, respectively. Both metals (V and Ni) have a bimodal distribution pattern, showing a little "hump" in the vacuum gas oil (VGO) boiling range. Direct measurement of V and Ni distribution by using high-temperature GC-SimDis with inductively coupled plasma mass spectrometry detection provides a close look at metal distribution in the deep-cut VGO boiling range (see Figure 1.21).

Literature data on oxygen concentration in petroleum are scarce, mainly because of low oxygen content in most crude oils and also because of difficulty in obtaining reliable results. Oxidation of the sample may further compromise the analysis.

1.7   CRUDE OIL MOLECULAR COMPOSITION AS A FUNCTION OF AEBP—THE CONTINUITY MODEL

Petroleum is a complex continuous mixture of homologous series of hydrocarbons and hetero-compounds, spanning a broad molecular weight and carbon number ranges. The increased complexity arises from the progression of initially narrow molecular weight (carbon number) ranges for low-boiling fractions, consisting primarily of hydrocarbons, to wider molecular weight (carbon number) ranges for high-boiling and nondistillable fractions, containing high concentrations of hetero-elements (S, N, O, V, and Ni).

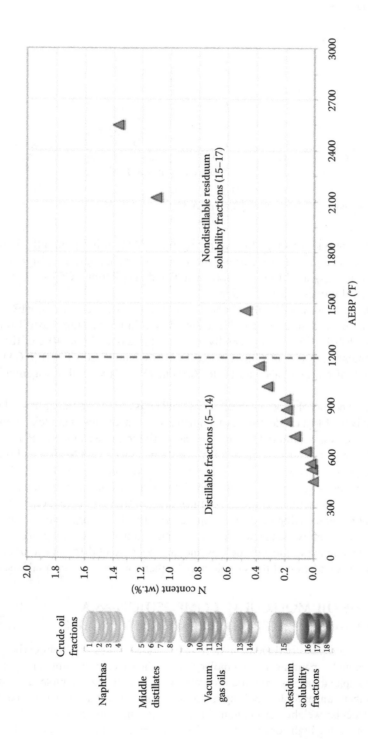

FIGURE 1.15 Nitrogen content in heavy crude oil fractions.

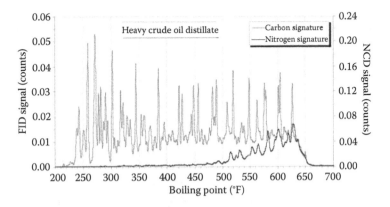

FIGURE 1.16   Nitrogen distribution in the middle distillate boiling range.

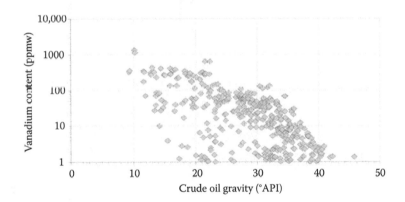

FIGURE 1.17   Vanadium content in different crude oils.

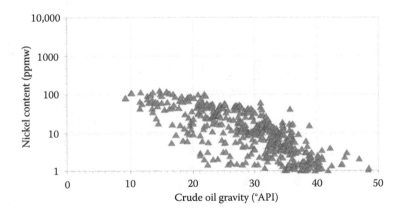

FIGURE 1.18   Nickel content in different crude oils.

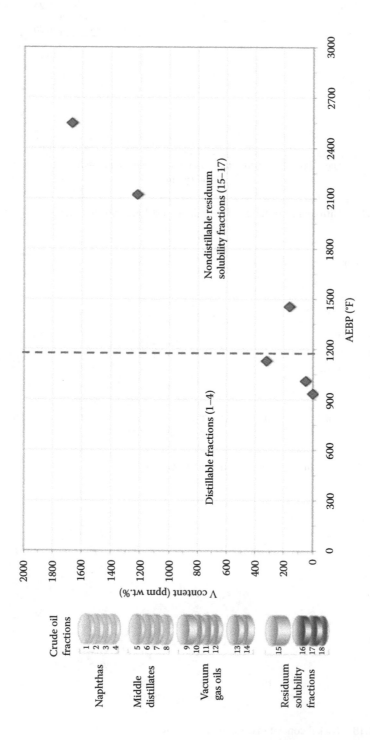

FIGURE 1.19 Vanadium contents in heavy crude oil fractions.

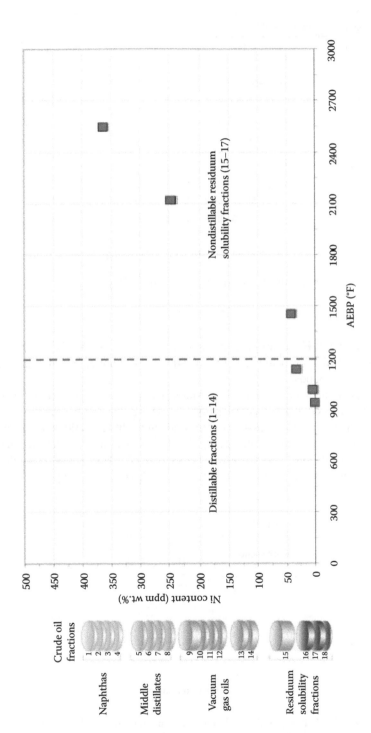

FIGURE 1.20 Nickel contents in heavy crude oil fractions.

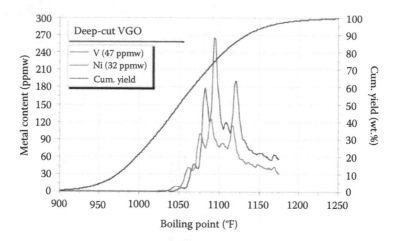

FIGURE 1.21 Vanadium and nickel distribution profiles in deep-cut vacuum gas oil.

All crude oil components can be described by the following general formula:

$$C_nH_{2n+Z}X$$

where C—carbon; H—hydrogen; X—hetero-elements (S, N, O, V, Ni); $Z = -2(R+DB-1)$, (hydrogen deficiency value), where R—number of rings and DB—number of double bonds.

The composition of a low-boiling crude naphtha fraction is a proverbial "tip of the iceberg" as compared with the composition of high-boiling and nondistillable crude oil fractions. The number of possible compounds in naphtha fractions increases rapidly with increasing boiling point, making the determination of the detailed naphtha composition a challenge. This is shown in Figure 1.22. There are 74 possible individual compounds, spanning the narrow C5–C8 carbon number range.

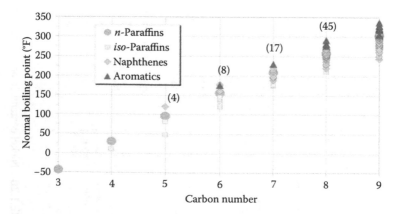

FIGURE 1.22 Number of possible individual compounds in naphtha fractions increases with increasing carbon number.

The number of possible compounds increases rapidly with increasing carbon number, making the analysis of high-boiling fractions in terms of individual compounds a formidable challenge. The number of possible isomers for paraffins alone (see Figure 1.23) increases exponentially with increasing carbon number (and increasing boiling point).

The diversity of compound types, each involving numerous homologous series, increases with increasing boiling point. The carbon number for each homologous series also increases with increasing boiling point. Furthermore, each individual carbon number homologue may be represented by a large number of possible isomers, resulting in very low relative concentrations of individual compounds. The immense complexity of crude oil composition, and in particular the composition of high-boiling and nondistillable "heavy" fractions, makes compositional characterization in terms of individual compounds (isomers) virtually impossible.

Great progress was made in recent years by Alan G. Marshall's research group [7–11], using ultrahigh-resolution Fourier transform ion cyclotron resonance mass spectrometry (FT-ICR MS) to provide detailed characterization of high-boiling and nondistillable petroleum fractions.

FT-ICR MS allows detailed characterization of complex petroleum fractions at the level of elemental composition assignment, providing unambiguous molecular formula assignment for each of the thousands of peaks in the mass spectrum. The authors reported that detailed compositional characterization of a heavy vacuum gas oil provided elemental compositions for >150,000 mass spectral peaks, which allowed for the calculation of double-bond equivalent (DBE) values for the identified homologous series.

Figure 1.24 illustrates the essential fact of petroleum composition: "Diverse compounds with similar molar masses cover a broad boiling range, and conversely, a narrow boiling range cut can contain a wide molar mass range" [1]. The chart serves as the foundation of the petroleum molecular composition continuity model. It provides

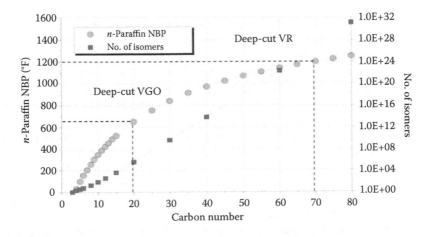

FIGURE 1.23 Number of possible paraffin isomers as a function of boiling point.

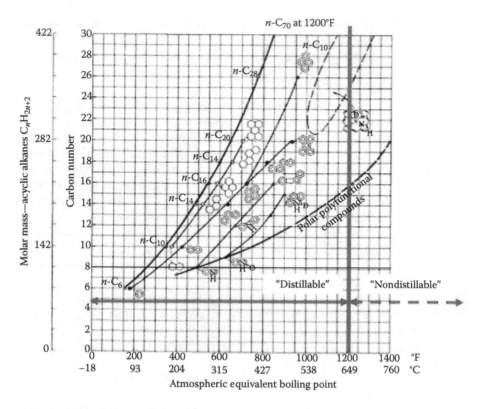

FIGURE 1.24 Molecular weight and boiling point relation.

an organizing principle that relates molecular structures and molecular weights to the AEBP. The chart reveals that at any given boiling point, hydrogen-rich paraffins have higher molar mass than the hydrogen-deficient polycyclic aromatic compounds and polar, polyfunctional hetero-compounds. The higher the boiling point, the wider the molar mass range.

The molecular weight (carbon number) distribution of heavy crude oil components has been a subject of many studies and controversies. Earlier reported results suggested that the great majority of heavy crude oil components, including asphaltenes, do not exceed a molecular weight of 2000 Da [1,2,6].

Molecular weight measurements for all distillable crude oil fractions and the first nondistillable residuum solubility fraction (SEF-1, #15) were performed using field ionization mass spectrometry (FIMS). The average molecular weights for all three nondistillable residuum solubility fractions, SEF-1, SEF-2, and SEF-3 (#15–17), were measured using VPO in pyridine. The results are shown in Figure 1.25. The VPO average molecular weight values for the three nondistillable residuum solubility fractions are erroneously high (particularly for SEF-2, #16 and SEF-3, #17) due to inter-molecular associations.

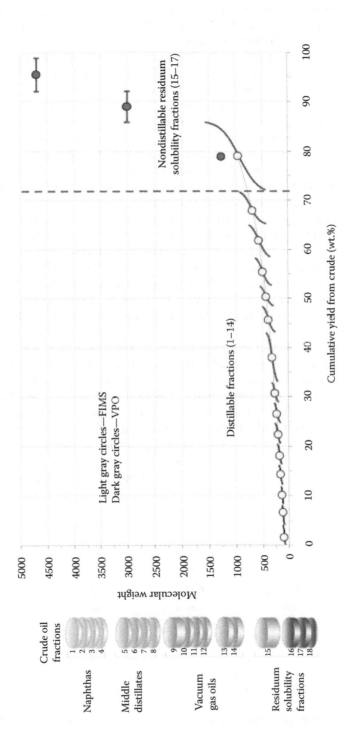

FIGURE 1.25 Molecular weight distribution of heavy crude oil fractions.

The hydrogen deficiency Z value in the general formula C_nH_{2n+Z} is a function of the number of rings and double bonds in the molecular structure. The hydrogen deficiency (or "aromaticity") can also be defined by the DBE value.

Figure 1.26 shows examples of fused aromatic ring–core structures and the relation between the hydrogen deficiency (Z and DBE) values and demand for hydrogen to hydrogenate (saturate) aromatic ring–cores without ring opening.

Figure 1.27 shows changes of the average Z values as a function of AEBP. The average Z values for all of the example heavy crude oil fractions were calculated using the average molecular weight values and carbon and hydrogen contents.

The hydrogen deficiency ("aromaticity") affects both the API gravity and the propensity to form coke. Figure 1.28 illustrates changes in micro carbon residue (MCR) with increasing AEBP. The very high MCR values for nondistillable residuum solubility fractions suggest a high degree of aromaticity.

The Z values for the nondistillable residuum solubility fractions are erroneously low due to the inflated, erroneously high VPO average molecular weights. To illustrate the effect of molecular weight on the calculated average Z value, three different average molecular weight values of 1000, 1500, and 2000 Da were used to calculate the average Z value for the whole nondistillable residuum. Figure 1.29 shows variations of the average Z value with changing average molecular weight. The plot also illustrates the relation between hydrogen deficiency Z value and API gravity.

The results suggest that the average molecular weight of approximately 1000 Da and the corresponding average Z value of approximately -48 best represent the nondistillable residuum. These results also suggest that the compositional continuity for the nondistillable residuum solubility fractions (SEF-1, SEF-2, and SEF-3) is determined mainly by higher hydrogen deficiency of the molecular structures (higher "aromaticity") rather than by higher molecular weight. Figures 1.30 and 1.31 provide a graphic illustration of the concept of the petroleum composition continuity model.

All model predictions have been recently confirmed for the full range of crude oil components [7–11].

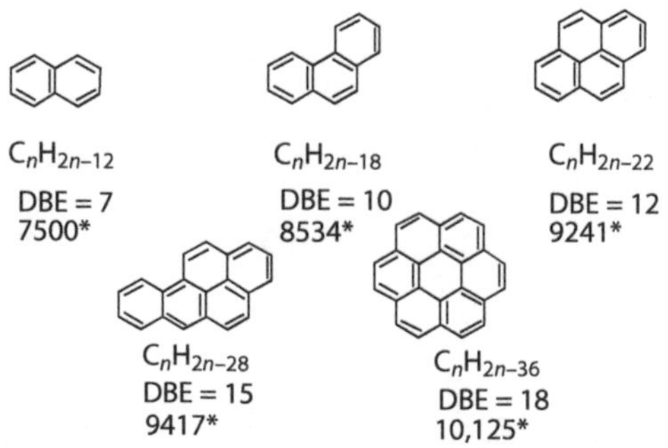

$$C_nH_{2n-12}$$
DBE = 7
7500*

$$C_nH_{2n-18}$$
DBE = 10
8534*

$$C_nH_{2n-22}$$
DBE = 12
9241*

$$C_nH_{2n-28}$$
DBE = 15
9417*

$$C_nH_{2n-36}$$
DBE = 18
10,125*

FIGURE 1.26 Hydrogen deficiency values and demand for hydrogen in SCF H_2/Bbl (asterisks).

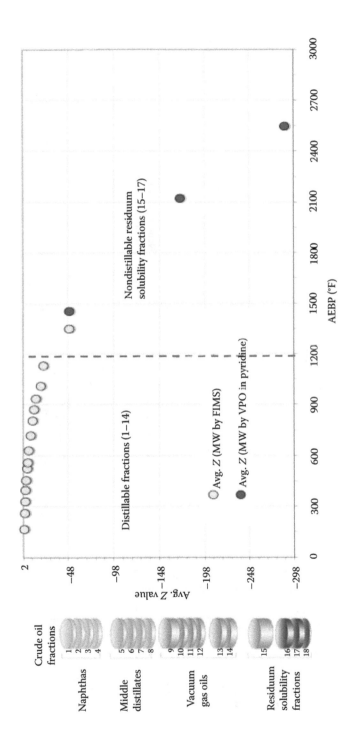

FIGURE 1.27 Average hydrogen deficiency Z values as a function of AEBP.

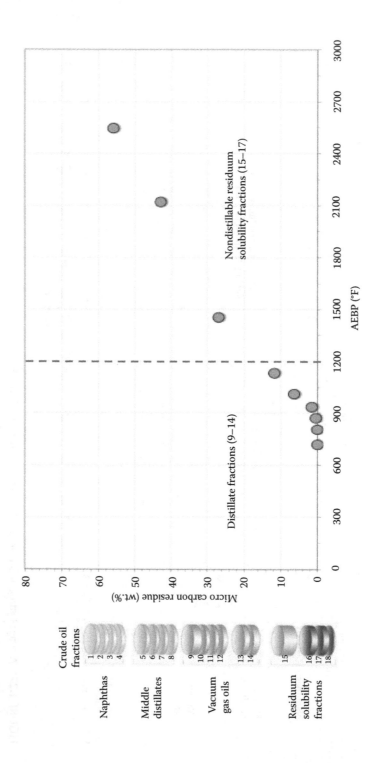

FIGURE 1.28 Micro carbon residue (MCR) as a function of AEBP.

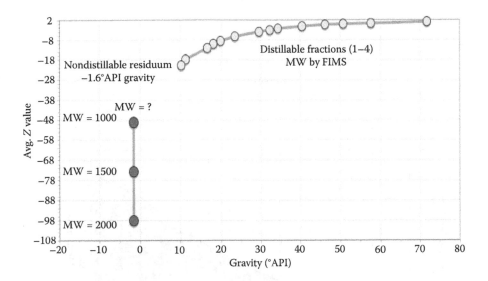

FIGURE 1.29 Variations of hydrogen deficiency Z value with changing average molecular weight values for the nondistillable residuum.

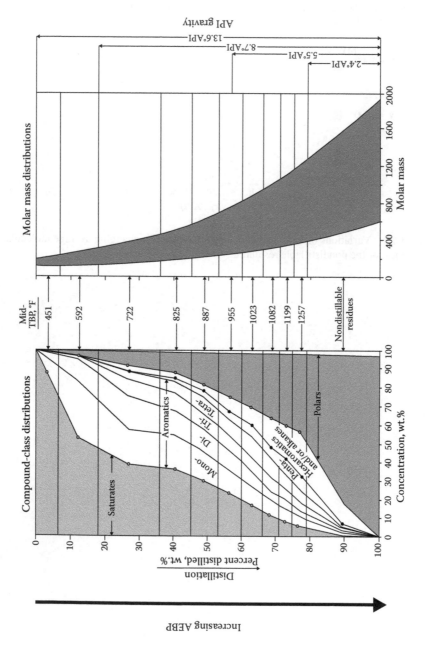

FIGURE 1.30 Compound-class and molecular weight distributions.

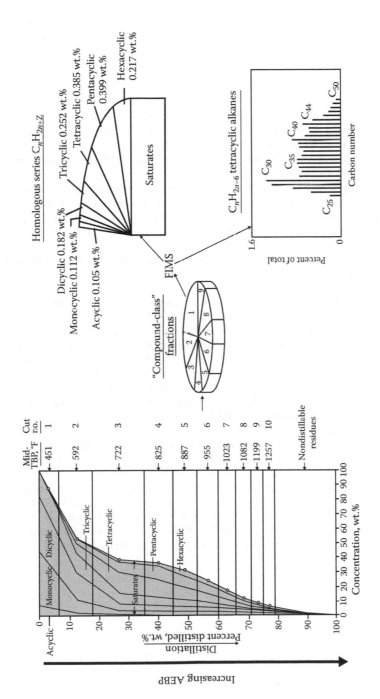

FIGURE 1.31　Homologous series and carbon number distributions.

ACKNOWLEDGMENTS

The author would like to thank Drs. Carl Rechsteiner and César Ovalles from Chevron Energy Technology Company for reading the manuscript and providing helpful comments.

REFERENCES

1. M. M. Boduszynski, "Composition of heavy petroleums. 1. Molecular weight, hydrogen deficiency, and heteroatom concentration as a function of atmospheric equivalent boiling point up to 1400°F (760°C)," *Energy & Fuels*, 1987, *1*, 2–11.
2. M. M. Boduszynski, "Composition of heavy petroleums. 2. Molecular characterization," *Energy & Fuels*, 1988, *2*, 597–613.
3. K. H. Altgelt and M. M. Boduszynski, "Composition of heavy petroleums. 3. An improved boiling point–molecular weight relation," *Energy & Fuels*, 1992, *6*, 68–72.
4. M. M. Boduszynski and K. H. Altgelt, "Composition of heavy petroleums. 4. Significance of the extended atmospheric equivalent boiling point (AEBP) scale," *Energy & Fuels*, 1992, *6*, 72–76.
5. M. M. Boduszynski, J. F. McKay and D. R. Latham, "Asphaltenes, where are you?," *Proceedings of the Association of Asphalt Paving Technologists*, Louisville, KY, February 18–20, 1980, vol. 49, pp. 123–143.
6. K. H. Altgelt and M. M. Boduszynski, *Composition and Analysis of Heavy Petroleum Fractions*, Marcel Dekker Inc., New York, 1994, 495 p.
7. A. M. McKenna, J. M. Purcell, R. P. Rodgers and A. G. Marshall, "Heavy petroleum composition. 1. Exhaustive compositional analysis of Athabasca bitumen HVGO distillates by Fourier transform ion cyclotron resonance mass spectrometry: A definitive test of the Boduszynski model," *Energy & Fuels*, 2010, *24* (5), 2929–2938.
8. A. M. McKenna, G. T. Blakney, F. Xian, P. B. Glaser, R. P. Rodgers and A. G. Marshall, "Heavy petroleum composition. 2. Progression of the Boduszynski model to the limit of distillation by ultrahigh-resolution FT-ICR mass spectrometry," *Energy & Fuels*, 2010, *24* (5), 2939–2946.
9. A. M. McKenna, L. J. Donald, J. E. Fitzsimmons, P. Juyal, V. Spicer, K. G. Standing, A. G. Marshall and R. P. Rodgers, "Heavy petroleum composition. 3. Asphaltene aggregation," *Energy & Fuels*, 2013, *27* (3), 1246–1256.
10. A. M. McKenna, A. G. Marshall and R. P. Rodgers, "Heavy petroleum composition. 4. Asphaltene compositional space," *Energy & Fuels*, 2013, *27* (3), 1257–1267.
11. D. C. Podgorski, Y. E. Corillo, L. Nyadong, V. V. Lobodin, B. J. Bythel, W. K. Robbins, A. M. McKenna, A. G. Marshall and R. P. Rodgers, "Heavy petroleum composition. 5. Compositional and structural continuum of petroleum revealed," *Energy & Fuels*, 2013, *27* (3), 1268–1276.

2 Process and Laboratory Sampling for Analytical Systems
Similarities and Subtle Differences

William M. Cost

CONTENTS

CONTEXT

Collection of good samples is a critical step before measuring crude oil properties. A modular system is described which

- Provides acquisition of samples at the point of interest
- Provides control to ensure representative sampling
- Is compatible with use both in laboratories and at the process line

ABSTRACT

The objective of any analytical measurement, especially for tremendously complex materials such as crude oil and its fractions, is to accurately represent the measured parameter. The first step in the measurement process is to obtain a representative sample of the material of interest. A further challenge is to obtain

31

the sample from a production or process environment that is representative of the material being sampled while the sample composition continuously varies.

Conventional tube and valve sample conditioning systems are widely used to obtain samples for in-the-field measurements or for later laboratory measurements. The physical size of such systems makes them impractical for laboratory use.

A recent miniaturized, modular sample conditioning system (ANSI/ISA SP 76.00102, sometimes called NeSSI—New Sampling/Sensor Initiative) bridges this gap. NeSSI systems have found increasing use in both the laboratory and process environments with features such as ease of use, simplicity in design, easy maintenance, and the ability to bring advanced analytical measurements to the point of interest, ideally mated to flowing systems.

This chapter discusses the application of such modular systems for the oil and petrochemical industries.

2.1   BACKGROUND

For operations in chemical manufacturing, refining, pharmaceutical, food and beverage, and biotechnology industries, as well as in municipalities, continuous monitoring of process stream chemistries or physical states is critical in sustaining production activities. Analytical systems typically employed for analyzing key component(s) in process fluids (liquids or gases) are of a wide variety dependent on stream matrix and component chemical structure. The currently used technologies span a wide range, with gas chromatography being the leading technique for high-level critical process measurement. In process control architectures, sampling and analysis are almost always supported or measured against a plant laboratory. Although laboratory measurements are completed under ideal ("ideal" is used here to describe a consistent environmental condition) conditions, the sample and how it was taken is critically important. In this regard, the process analyzer and laboratory instruments are similar. Both sampling approaches (process and laboratory applications) require a sample that is conditioned properly for generation of reliable data transmitted to the plant operator or controls engineer.

There are basic fundamentals of sample conditioning that must be maintained or controlled in both sampling operational schemes. As a general rule, the fundamental parameters of pressure, flow, and temperature form the framework of all sampling techniques and their proper control will limit the effectiveness of any analytical measurement. Each of these parameters will have varying levels of importance in each application based on stream chemistries encountered and the type(s) of analytical measurement implemented. Also, the conditioning location (process or laboratory environment) will have a direct impact on how samples are conditioned. For instance, process analytics will require proper temperature control for applications where sample condensation or vaporization is critical to the presentation to analytical instrumentation. Also, in process environments, flow is the most critical parameter owing to the direct impact of correlating sample flow to process changes during dynamic operation. Flow is a fundamental parameter for separation techniques such as gas or liquid chromatography but is also critical in providing consistent sample to a process analytical system. Therefore, in a process application, flow could be considered the baseline requirement of the physical parameters mentioned previously. A key difference in flow control or monitoring in process and laboratory applications is the flow medium itself. Process measurements are taken in a dynamic matrix under the constant flow of medium to be measured. For example, a process gas chromatograph (GC) will have samples presented to the analyzer on a continual basis and the sample injected into the analyzer is a process stream sample. In a laboratory environment, the same measurement may be conducted by injecting the process sample with a different balance of solvent in a static process. Static here is referring strictly to the sample itself. The sample to be measured is not a part of the continuous flowing matrix. The parameter of flow is important in laboratory applications and specifically in the example given here referring to carrier gas flow. Without proper carrier gas flow control with a laboratory GC, poor separation would result (i.e., varying peak retention times, poor peak resolution, etc.). Therefore, flow for laboratory applications is, in most cases, referring to the medium transporting the components to be measured (i.e., gas chromatography carrier gas or liquid chromatography mobile phase[s]). Of course, this is not always the case and samples can be injected directly into a laboratory analyzer from a sample cylinder or similar device. However, to draw a distinction between process and laboratory applications, flow can be used as a basis for differentiation between the two environments.

Pressure reduction and control is required in many process monitoring applications to provide the analytical system(s) a sample within the pressure specifications of the analytical technique, maintain the sample phase (liquid or gas), or return the measured sample back to the process or a disposal location (flare). In laboratory applications, pressure and its criticality is in reference to the analytical device and sample phase. If the sample delivered to the laboratory is under positive pressure or vacuum, this should be maintained in presenting the sample to the instrument. This may present difficulty with laboratory instruments, as they generally require positive pressure to present sample to the measuring device. This is a significant difference between laboratory and process measurement. Process analyzer sampling systems must reduce pressure or supply motive force (sample pump or aspirator) to present a sample for analysis. In this case, pressure has a similar importance to flow in both sampling applications. Pressure and its control have an added criticality for analysis owing to its effect on the sample matrix. Depending on the sample, pressure can be used to assist vaporization or condensation of sample for monitoring. Also, pressure stability is of utmost importance for partial pressure sensitive analyzers such as O_2, H_2, and infrared. In process monitoring, continuous sample supply to the analytical system and the variable nature of process samples require hardware for pressure control. For laboratory systems, pressure control is also important but is typically done within the analytical device, or the analytical technique used is different than that administered in the field.

In some process and laboratory sampling applications where analyzers are used to measure oxygen, stack gases, hydrogen, etc., heating may be required to maintain the sample gas above the dew-point temperature, preventing sample condensation. Temperature and pressure control are also very important in applications that require vaporization of a liquid before measurement. Typically, laboratory analytical applications are conducted at consistent ambient temperatures; therefore, temperature effects are minimized in comparison with the process sampling application. However, in laboratory and process applications that require similar sampling architectures, the same pressure and temperature control attributes are required to effect efficient vaporization.

The hardware implemented to achieve proper sample conditioning (i.e., the above-mentioned control of key parameters) can vary in form factor but is largely composed of two separate form factors. The first and most common sampling system approach is a tubing and compression fitting architecture (see Figure 2.1). The second approach comprises modular sampling fittings and control components (ANSI/ISA

FIGURE 2.1 Traditional sampling system architecture commonly used for process analytical measurement.

FIGURE 2.2 Typical modular sampling system architecture for process analytical measurement.

SP76.00.02 system; Figure 2.2). The architecture of each sampling system type has specific benefits, and each may provide a sample that is delivered to the analytical device with a high level of consistency. The array of flow capabilities outlined within the ANSI standard provides a vast arrangement of control and application capabilities. An overview of flow fittings (i.e., substrates) is depicted in Figure 2.3a. The

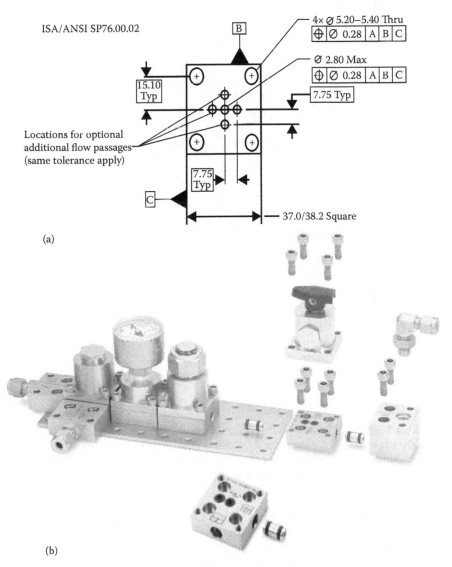

(a)

(b)

FIGURE 2.3 (a) Basic modular sampling system standard that forms the basis for modular sampling designs. (b) Parker Intraflow sampling system configuration demonstrating the basic design architecture. Parker's system uses a pegboard mounting geometry, 10/32 screws, and slip-fit connectors to assemble substrate fittings.

(Continued)

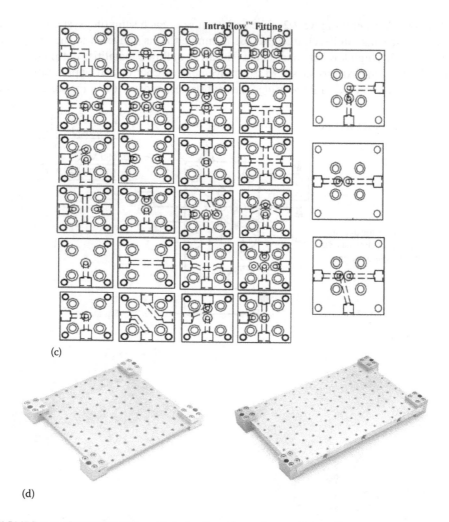

(c)

(d)

FIGURE 2.3 (CONTINUED)   (c) Parker Intraflow substrate fitting library. This is a partial list, as the total number of substrate possibilities is close to 100. (d) The left pegboard is 1/8″ (3.1 mm) 316 SS and the right pegboard is 1/2″ (12.7 mm) aluminum. The aluminum pegboard may be fabricated with milled ports for flowing coolant.

ability to machine different flow paths within a substrate component and the capability to accommodate higher-precision fluidic controls gives the end user unmatched control capability within a reduced sample system geometry.

The key to any successful analytical measurement, whether in the process or laboratory environment, is to eliminate sampling inconsistency. The focus of this chapter is to review the sampling techniques, traditional and modular, outlining the performance enhancements attainable by using a modular platform and the versatility it provides over compression, traditional tube-based systems.

2.2 MODULAR SAMPLING PLATFORM—WHY?

Sampling system designs have experienced minimal changes since the first attempts of conducting process analytical measurements. During this period, many techniques were applied and implemented for process sampling. However, in the late 20th century, a group of scientists and industry leaders formed a sampling system initiative sponsored by the Center for Process Analysis and Control at the University of Washington. This initiative was called NeSSI (New Sampling/Sensor Initiative). This approach was considered as an alternative method for reducing sample conditioning cost, both installation and the long-term cost of ownership. Although this was a paradigm shift in sampling systems, it would be required to follow the same basic sample conditioning fundamentals that had been successfully employed for decades. The baseline criteria for this new design must maintain the reliability and safety attributes of previous conventional designs but should bring value by increasing efficiency and flexibility while reducing undesirable maintenance downtime. The NeSSI design provided the engineers, technicians, and original equipment manufacturers (OEMs) of analyzers a consistent hardware template that would allow multiple supplier components or systems to be used within a single sample conditioning application. The basic design feature for modular sampling systems is outlined in Figure 2.3a through c. This is also known as the ISA/ANSI SP76.00.02 standard for modular sampling systems. From this standard, suppliers (Parker Hannifin Corporation, Swagelok, and Circor Tech) developed their own unique approach for sample system control from the base standard. For the discussions included herein, the Parker modular system will be used exclusively as the modular template (see Figure 2.3b). The array of flow capabilities outlined within the ANSI standard and provided by Parker Hannifin Corporation gives the end user a vast arrangement of control and application capabilities. Also, the pegboard mounting support for these systems gives an added advantage to mount other necessary hardware if needed. This will be discussed going forward, in detail, regarding hybrid sampling applications. An overview of flow fittings (i.e., substrates) is depicted in Figure 2.3c. Note: Other supplier systems could be inserted here as options for sample conditioning. The key point is to bring forward the attributes of modularity and how it can be used in process and laboratory environments.

2.3 MODULARITY FOR PROCESS SAMPLING APPLICATIONS—WHY?

The modular footprint is ideal for process locations where space limitations require a reduced sampling dimension. The modular aspect of sampling allows the end user (as long as the samples are conducive to modular hardware) to control pressure, flow, and temperature as they would in a traditional sampling system. The modular platform, specifically Parker Intraflow, provides a significant advantage over traditional designs that require temperature control. This system allows for both conductive heating and cooling. Given the higher thermal mass of the Intraflow system, temperature can be tightly controlled over a relatively broad range. From an economic perspective, a modular sampling system can also provide a temperature control advantage by requiring less power to maintain temperature owing to the dimensions of the physical system. Traditional compression tube sampling systems can also provide precise temperature control; however, this involves the use of insulated and electric traced sample tubing. This approach works well but can be difficult to maintain; it can also be difficult to perform routine tasks given the tubing insulation requirement.

Modular systems may provide high-level pressure and flow control comparable to traditional sampling systems by using readily available hardware from multiple manufacturers. For mass and volumetric flow control, hardware components are available from Brooks Instruments, Porter Instruments, Alicat Scientific, and others. Pressure control may also be optimized by using hardware available from multiple manufacturers (Alicat Scientific, Porter Instruments, GO, Verfilo, etc.). Modular systems provide another unique mechanical advantage to flow control by using orifice plates that can be mounted into the flow path with substrate fittings and are interchangeable (specifically Parker Intraflow). The use of an orifice plate for flow control is a lower-precision approach; however, it can be implemented easily in situations where their use is acceptable. The combination of high-level temperature control with precise pressure and flow control options provides the end user with a powerful platform to conduct process analytical measurement that is comparable to traditional compression-tube-style sampling conditioning systems. Also, the inherent hardware flexibility with modular systems and the built-in capability for system expansion for additional process measurement needs give modular sampling systems a hardware advantage for many process analytical applications. This is possible with the modular system platform since fittings are not connected by permanent compression-style connections (see Figure 2.3). Given the fluidic control interface with modular systems, maintenance functions may be completed by a small number of tools, and in most instances a single tool. This is the maintenance advantage realized by use of modular sampling systems in the process environment.

We have discussed sample control parameters required for analysis; however, to this point, no discussion has been given to the preparation of the sample matrix for analysis. The preparation referred to here is sample filtration (i.e., particulate in-line, particulate bypass, coalescing, etc.). The details of filtration types will not be discussed here; however, a brief discussion of sample filtration must be included in the context of sample systems analysis. Most samples are filtered before analysis in process analytical system applications (there is a small percentage of measurement that is conducted in situ in the process and does not include typical filtration methods). Modular systems are designed with the same filtration concepts. There are differences in sample volumes and flow rates in each sampling approach; however, the concept of filtration and where the different filtration methods are used are the same. Also, for demanding samples (i.e., heavily particulate-laden samples), there is another approach that modular systems may provide the end user. This approach may be called a hybrid configuration. In the hybrid configuration, traditional filtration methods may be employed, and a modular sampling control architecture is installed downstream of the filter (see Figure 2.4). This approach gives modular systems, which already have a wide range of application capability, an even greater range in process applications.

Along with the wide application range of modular systems (i.e., low-sulfur diesel to synthesis gas measurement), the modular architecture provides a great platform for

FIGURE 2.4   Hybrid sample system design using higher-flow filtration upstream of the Parker Intraflow modular sampling system.

standardization and subassembly design. Owing to the single tool assembly nature of modular systems and their component design, standard systems are easily duplicated and built on a large scale. The ease of assembly allows a single fabricator the capability to construct multiple systems in a short period of time. For the OEM supplier, standard systems can be installed within traditional architectures mentioned previously, with little effort. In today's world of low inventory and build to order accounting practices, standardized systems provide a supplier, and ultimately the end user, the capability to respond quickly to urgent sampling needs. A representation of a standardized sampling system schematic is given in Figure 2.5.

Standardized systems can also be used to assist in large analyzer project designs, and allow analyzer support personnel a system structure that provides a common

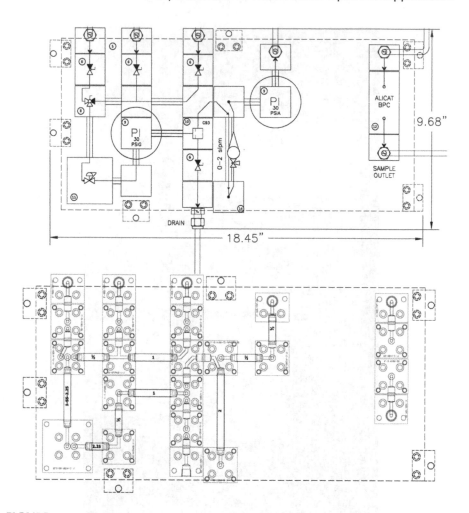

FIGURE 2.5   Top-level fluidic control and substrate-level representation.

support and troubleshooting template for maintenance tasks. This is critical in routine preventative maintenance and support efforts. Over the operational life of an analyzer system, this could translate into thousands of dollars. Modular systems may be constructed as subassemblies within an overall sample conditioning system, similar to a hybrid system. However, in a hybrid system, the modular hardware comprises the majority of the sampling system. A true modular subassembly would be hardware for fast-loop filtration, sample bypass, stream selection valves, permeation tube assemblies, pressure control manifolds, gas purification, and gas blending. All of these can be fabricated as a modular sampling subassembly. The benefit of these assemblies would be for drop-in replacement of components for field maintenance or as an addition to an existing sampling system. Examples of gas blending and fast loop filter assemblies are outlined in Figure 2.6a and b.

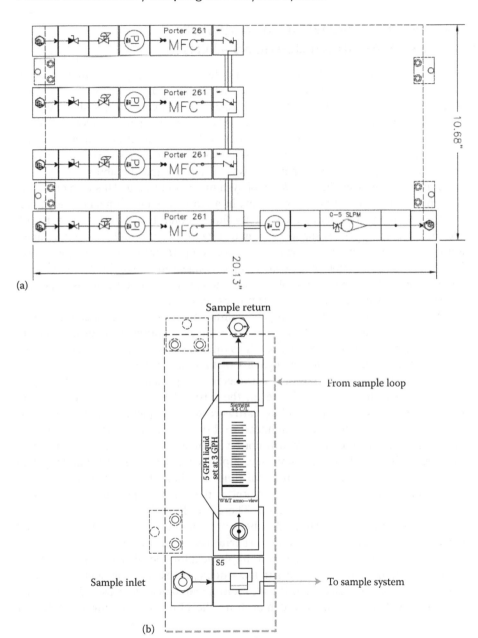

FIGURE 2.6 (a) Modular gas blending assembly design (Parker Hannifin Corporation). (b) Fast-loop filtration assembly.

2.4 MODULAR SAMPLING SYSTEMS FOR LABORATORY APPLICATIONS—WHY?

As mentioned previously, the attributes of modular sampling systems are also applicable to laboratory systems and testing. In fact, a laboratory environment may have more spacing limitations for conditioning systems and often needs a system that will run in a batch sampling or reaction format. Also, in research and development environments, a system that is conducive to routine flow and hardware changes is preferred. A modular architecture meets these needs quite well. In the process sampling discussion, sample stream temperature was discussed and the ability of modular systems to conductively heat and/or cool systems was outlined. This is another significant factor for use of modular systems in a laboratory environment. The Parker system, owing to its mounting interface for fluidic control hardware (i.e., pegboard; see Figure 2.3d), provides support for small reactor systems and localized temperature control that may be needed for reactor temperature control, sample vaporization, or the prevention of gaseous sample condensation, similar to the process environment. The ability to control pressure, flow, and temperature are as important in the laboratory environment as in the process environment.

Many laboratory applications are low-volume and -flow, single-batch techniques that do not run continuously. Of course, this is an assumption and does not apply to all the possible laboratory applications. The intent here is to outline capabilities and discuss specific examples for the benefits of modular systems. Thus far, we have discussed the three major physical parameters for presenting consistent samples to analytical instrumentation. However, the basis of the NeSSI initiative was to include the use of modular sampling hardware with high-level analytical systems and digital bus communication architecture. For the present context, a discussion of communications architecture will not be included. It is important to recognize the value of being able to communicate sample system health (smart sample system approach) and analytical data on a real-time, continual basis. As many analyzer engineers, technicians, and researchers know, the data transmitted is only as good as the sample that is being measured. Therefore, how the sample is conditioned is of utmost importance. Although laboratory samples are of lower volume and batch in nature, all of the sampling criteria discussed are relevant. The laboratory environment allows the end user to implement high-level analytics and sensors more freely than their counterparts in hazardous process environments. Thus, as a few examples are discussed in this text, multiple analytical devices and supporting equipment will be presented to demonstrate the ultimate sampling flexibility of modular systems.

2.4.1 MODULAR LABORATORY SYSTEM EXAMPLE #1— FERMENTATION OFF-GAS MEASUREMENT

As discussed earlier, the use of modular systems provides numerous benefits for laboratory-based analytical applications. The first system review was for a fermentation off-gas measurement. The key sample specifications were that the gas was at atmospheric pressure, ambient temperature, and consisted of multiple streams.

The challenge was twofold: first how would the sample be delivered to the sampling system and control flow across multiple streams. Second, how would the analyzer be calibrated? Both of these issues are very important factors to consider. The first issue was addressed by implementing a sampling pump (Air Dimensions Inc.) that could be mounted to the sampling system. Note: A common issue could be saturated vapor from the fermenter. However, the system operated at ambient temperature and the volume reacted minimized any moisture carryover. Also, the sampling system was heated to 55°C above the sample gas dew point. The second issue was how to deliver calibration gas to the GC and reference atmosphere for consistent sample injection volume. This was accomplished by discrete outputs located in the analyzer and transmitted through an I/O module interface via the GC method. The calibration gas was provided by a Kin-Tek permeation device. At controlled flow and temperature, a consistent calibration gas was delivered to the GC. The end result of these sampling requirements resulted in the system represented in Figure 2.7a and b.

2.4.2 Modular Laboratory System Example #2—Gas Calibration Systems

The need for blended gases to calibrate process laboratory GC or other process laboratory instrumentation is vital in applications where laboratories are required to validate process measurement. Also, gas blending is a needed where expensive blended gases are used in high volumes. The use of a blending system may minimize the cost of such calibration gases by blending multiple pure gases for a desired component blend concentration. The examples outlined here are blending multiple gases from cylinders and using a permeation device for blending the output gas to the desired concentration. The first system schematic is presented in Figure 2.8. This system used mass flow controllers, pressure regulation, and a permeation device for blending. Another gas blending example used both volumetric (using back-pressure regulation) and electronic flow control devices to control gas flow and delivery to a laboratory GC system. The system is represented in Figure 2.9.

2.4.3 Modular Laboratory System Example #3— Microreactor Monitoring Systems

Modular sampling systems not only control pressure, flow, and temperature but they also may be used for continuous monitoring of a reaction. As mentioned previously, these systems may incorporate analytical measuring devices into the sampling architecture. The example outlined here was used for continuous monitoring of a chemical reaction process and measured output to determine reaction progress. The basic sampling system design practice of controlling pressure, flow, and temperature was addressed, as well as using pressure and flow transmission devices to tightly control reactor input and output. The system represented was also unique in that it used a cooling process to reduce heat output from an exothermic chemical reaction and control input temperature of the reactants (see Figure 2.10). Results of the testing with this system demonstrated the viability of using modular hardware in a microreactor application during continuous operation. This system represents a "smart"

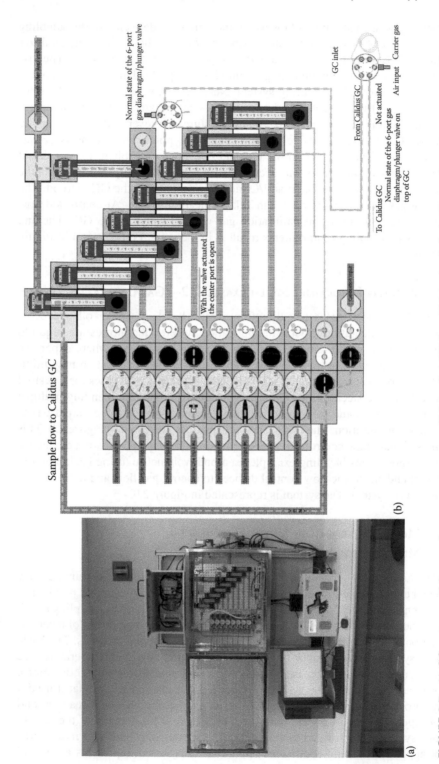

FIGURE 2.7 (a) GC system used for sample measurement was a Calidus fast GC from Falcon Analytical. (Photo of sampling system courtesy of Falcon Analytical.) (b) Flow schematic representation of the sampling system. (Schematic courtesy of Falcon Analytical.)

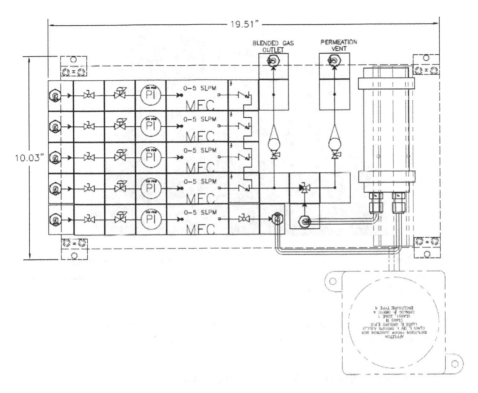

FIGURE 2.8 Process laboratory gas calibration system using a permeation module and cylinder gases.

FIGURE 2.9 This is an electronic and manual gas blending application. The manual or volumetric flow control system can be seen on the left side and the electronic system is on the right side of the photograph. (Systems courtesy of Chevron Corporation, Richmond, CA.)

FIGURE 2.10 This sample system application and testing was completed by Brian Marquardt, APL University of Washington. (Photo courtesy of Brian and his analytical team at the Applied Physics Laboratory.)

system by using sophisticated measurement techniques and compiling data in real time to monitor reaction status and sample system health.

2.4.4 MODULAR SAMPLING SYSTEM EXAMPLE #4— AT-LINE LABORATORY APPLICATIONS

In some instances, end users have the availability of laboratory-grade analytical instrumentation that could be used in a semibatch application. This is similar to a pilot plant scenario but generally do not have the hazardous area restrictions that eliminate many laboratory-based GC systems from operating in these areas. As modular systems are used for continuous on-line analyzer and laboratory-based applications, they are also used for what is known as "at-line" measurements. Once again, due to a modular system's smaller dimensional geometry, they work well in applications that require a reduced footprint. The example reviewed in this instance is a laboratory GC application that required the sample to be vaporized before sample injection. A VICI electronic injector was used to deliver the sample to the GC injection port, and a Parker vaporizing regulator was used to vaporize the sample before delivery to the VICI valve (see Figure 2.11). Another at-line approach is to use a three-dimensional sampling system (Astute USA and EIF) to extract, filter, stream, select, and return the sample back to the process. This is a unique approach and has been used with success in different applications. The concept is to minimize worker exposure, minimize sample delivery lag time, and reduce the need for expensive environmental controlled shelters for process analytical instrumentation. Concepts like "at-the-pipe" measurement are in the early stages of development and acceptance but have great promise going forward from process monitoring. The examples given are a very small representation of the variety of applications using modular sampling system architecture. The intent was to introduce the technology

FIGURE 2.11 Sample system with vaporizing regulator and electric actuated VICI selector valve.

and expand on the breadth of applications that can be addressed by using modular components. From these application descriptions, it is obvious that modular systems perform well and may be used in a vast array of sampling applications. Inherent advantages are reduced footprint, flexibility, subassembly design, standardization, analytical instrument interfacing, improved temperature control, and capability to be implemented with ease in the field, process laboratory, R&D laboratory, and at-line. The modular architecture is not a new concept but rather a tried-and-true method for conditioning and analyzing sample streams in multiple locations. As a former end user of sampling systems, I believe it is beneficial to be aware of multiple options for completing a task. Being properly informed of hardware capabilities is a good first step in successfully applying the knowledge gained.

Section II

Water Analysis

3 Advances in Oil-in-Water Monitoring Technology

Darrell L. Gallup

CONTENTS

CONTEXT

Since oil production produces almost an order of magnitude more water than oil, accurate measurement of oil and water is critical to

- Optimize the extraction of oil from the water
- Minimize the environmental impacts of wastewater disposal
- Minimize damage to disposal wells

ABSTRACT

The concentrations of dispersed and dissolved oils, solids, and additive chemicals in produced or injected water are typically restricted under environmental guidelines or operational requirements. The former is important when discharging water to the environment, while the latter is needed to avoid disposal well formation damage. Returning excess petroleum product to the reservoir in injection water is economically undesirable. The quality of disposed or injected water is monitored using a variety of methods. Continuous monitoring of chemical constituents in water is a common practice in industry, and is usually favored over manual monitoring. Technology in continuous monitoring of oils, solids, treatment chemicals, and naturally occurring environmentally sensitive constituents is advancing, particularly in detection applications. Real-time monitoring of water in the presence of optical foulants is accomplished with ultrasonic cleaning methods. Technological advances in on-line and in-line monitoring of oily or additive components in disposed water are presented and discussed.

3.1 INTRODUCTION

The concentrations of dispersed and dissolved oil, solids, and additive chemicals in produced or injected water are typically restricted under environmental regulations or operational requirements. The former is especially important when discharging water to the environment, while the latter must be considered to avoid disposal well formation damage. Returning excess petroleum product to the reservoir in injection water is economically undesirable. Disposed or injected water quality is monitored by using a variety of methods. Continuous monitoring of chemical constituents in water is a common practice in industry, and is usually favored over manual monitoring methods.

Technology applied in continuous monitoring of oil, solids, treatment chemicals, and naturally occurring environmentally sensitive constituents is advancing, particularly in detection applications. In recent years, we have surveyed manufacturers of oil-in-water (OiW) monitors in an effort to find highly reliable, accurate, and precise instruments for field deployment. More than a dozen monitors were included in the survey, which consisted of obtaining information from suppliers, visiting some manufacturer's facilities, and observing the performance of instruments in the field.

The vast majority of the continuous, on-line/in-line monitors that were surveyed utilize ultraviolet (UV) fluorescence of organic compounds for the detection and quantification of OiW. A few use other detection methods, such as gas chromatography. The first continuous, on-line OiW monitors developed utilized UV fluorescence detection through "free-falling" water in a glass column. The purpose of this design, which became very popular in industry, was to keep the water from contacting the glass in an effort to avoid fouling of the glass windows. Monitors based on this technology allowed continuous or sometimes intermittent "real-time" monitoring for operations to observe the behavior of water treatment processes and to avoid discharging excess oil to the environment. Oil sheen formation on water could be avoided by the fast response and feedback to operators of the water treatment systems.

Some drawbacks to this early technology included fouling of the optics and other pieces of equipment with emulsions or solids. Occasionally, the water splashed onto the glass windows, reducing the UV light intensity. Unless operators realized that fouling had occurred, inaccurate measurements could accrue. Especially where emulsions and solids were still present in water exiting the treatment system, operators and technicians were required to maintain the instruments with high frequency. In one example, an offshore field used a precipitation process in the water treatment system to prevent the release of heavy metals into the ocean. The precipitates were quite "sticky," especially to glass windows. Platform operators were required to nearly continuously maintain the OiW monitors. In another example, an offshore field produced a very stable oil–water emulsion that would often splash onto the optics, thereby confusing the operators when the monitor would produce low readings of OiW; however, oil sheens were observed at the water discharge pipe. As a result of some of these maintenance problems, technicians and operators would simply tire of cleaning glass windows, and decisions were made to turn off the OiW monitors and rely solely on "grab" sample analyses. Furthermore, when the operators and technicians were trying to maintain the monitors, they frequently would lose their ability to calibrate the instruments.

The result of these problems encountered with the first-generation technology was a search for alternatives that might be easier to maintain and operate, especially in remote locations. The survey for second-generation technology conducted in the

TABLE 3.1

Some Vendors of OiW Analyzers and Analyzer Principle

Vendor	Analyzer Principle
Advanced Sensors	Laser-induced fluorescence spectroscopy
Analytical Systems International	Organic stripping and detection
Arjay Engineering	UV fluorescence spectroscopy
Canty Inflow	Microscopic imaging
Mirormax	Acoustic reflection/absorption
Proanalysis Argus	Laser-induced fluorescence spectroscopy
Systektum Kontavisor	Laser-induced fluorescence spectroscopy

present study focused on "self-cleaning" monitors that would reduce operator maintenance, excessive calibration, and reverting to frequent grab sample analysis to not only ensure discharge water quality but also to understand the behavior of water treatment processes. Table 3.1 lists the vendors and types of analyzers surveyed. The desired features of new OiW monitors would be alarms and trend analyses, and not necessarily reporting to governmental agencies of regulation compliance. Improved monitors would be primarily used as "early warning" systems of problems or upsets with water treatment systems to prevent excess oil discharge and sheens on water bodies.

3.2 TECHNOLOGY SELECTION

Our survey discovered several new monitors that could potentially monitor OiW in real time in the presence of fouling agents that incorporated ultrasonic cleaning methods. The monitors could be used on-line or in-line to detect oily or additive components in the water that fluoresce. The new monitors discovered in the survey continued to use the original laser-induced UV fluorescence technique for detection of hydrocarbons.

We found one particular technology that met all of the desired features of improvements. The transducer sensor head used combined optical and ultrasonic components. In this monitor, water passes through a measurement chamber, which includes a small sapphire window used for measurement and a larger sapphire window for viewing the inside of the chamber. The laser light passes through the smaller sapphire window to excite the compounds in the water containing C–C double bonds. The emitted fluorescence is captured via optical fiber light guides and taken into an optical filter and photomultiplier tube. The selected optical filter depends on wavelength properties in the water. In another feature, an optical UV spectrometer is added. The ultrasonic transducer in the instruments performs two functions: (i) it cleans the sapphire windows ensuring that measurements do not deteriorate due to fouling, and (ii) it periodically homogenizes the sample to correct for oil droplet size. Figure 3.1 diagrammatically shows the chamber in the monitor.

The built-in UV spectrometer scans the water every 10 min to provide detailed analysis of the water. Any interfering species in the sample may be identified, and

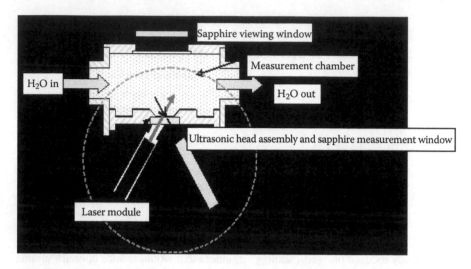

FIGURE 3.1 OiW schematic.

the wavelength range may be adjusted to eliminate the interference. Figure 3.2 shows that a chemical used in water treatment is exacerbating fluorescence beyond the oil signal at about 460 nm. By shifting the wavelength to above about 600 nm, the interference is essentially subtracted. Although additive chemicals may be evaluated in the laboratory before installation of a monitor, the built-in spectrometer facilitates the evaluation and analysis of new chemicals/additives being introduced into the process water.

The spectrometer detects the specific fingerprint of the oil present in the water, and should the oil in the water change in composition, the spectrometer will generate

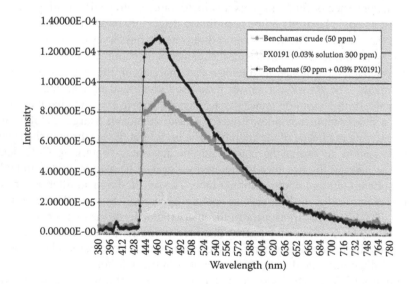

FIGURE 3.2 Spectrophotometric scan of water, chemical, and chemically-treated water.

a water quality alarm. The spectrometer can be used to identify the specific type of oil present in the water, and once identified the analyzer can select the appropriate measurement calibration for the oil to maintain measurement accuracy. Should a process upset occur, the spectrometer data logs can be retrospectively reviewed to assist in identifying the cause of the process upset, for example, the use/misuse of chemicals or the oil type present in the water.

The dynamic range of most OiW monitors is limited to about 500 ppm OiW or a maximum of 1000 ppm total petroleum hydrocarbons. A further advancement made to this preferred OiW monitor technology is a direct fluorescent dynamic range of ~20,000 ppm without any extrapolation. Above ~20,000 ppm OiW, the fluorescence signal flatlines. The flatlining in itself is a new development. Traditionally, as oil concentrations increase, the fluorescence flatlines and then decreases, which can confuse an instrument into believing that the oil content is falling when it is actually still increasing. With this extended range feature, it is possible to monitor not only water discharge/reinjection quality but also the efficiency and process dynamics of water treatment equipment (where concentrations are <2%), such as separators, hydrocyclones, filters, and flotation devices.

3.2.1 OPERATION

Upon installing the new, advanced OiW monitors, it is important to understand and correlate the concentrations reported by the instrument to the concentrations reported to regulatory bodies that often require grab sample and analysis by gas chromatography, gravimetry, or extraction/infrared (IR) methods (e.g., EPA 413, 415, and 418). An example of the correlation between output concentrations from the new, improved OiW monitor and the laboratory grab sample extraction/IR method is shown in Figure 3.3. In this example, remarkable agreement between the OiW monitor and the extraction/IR method is attained.

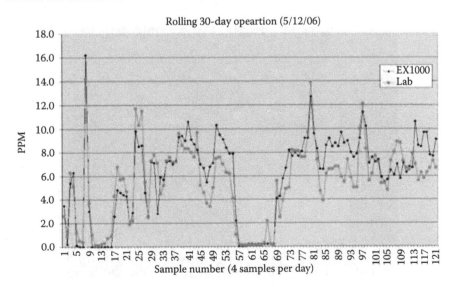

FIGURE 3.3 Comparison of OiW instrument and extraction/IR of "grab" samples.

3.2.2 TECHNOLOGICAL ADVANCES

Since installing a number of these preferred OiW monitors in various fields worldwide, the manufacturer has developed some additional capabilities that we plan to examine. One advancement is a microscopic tool addition with the OiW monitors to determine oil droplet and particle size. The analyzer captures images of the sample water and uses software to detect the oil and solid particles in the water. The software assesses each particle and droplet size, and then produces measurement data. The microscopy unit also has an automatic configuration capability that requires no routine user configuration that could lead to frequent operator adjustment. By combining both the fluorescent measurement and video microscopy capability in the

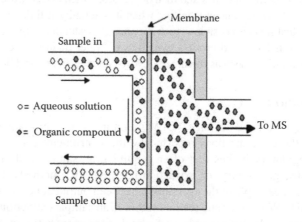

FIGURE 3.4 Principles of MIMS.

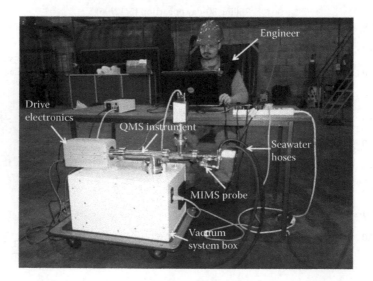

FIGURE 3.5 MIMS addition to OiW monitor prototype.

same analyzer (including the self-cleaning feature), it is possible to achieve the OiW measurement accuracy of fluorescence, and the oil droplet and solid particle sizing capabilities of video microscopy. The latter is offered as a stand-alone monitor by several suppliers.

Another advancement that is under development is the addition of a membrane injection mass spectrometer (MIMS) to the OiW monitor (see Figure 3.4). While monitors normally measure the effects of oil present in the water using techniques such as fluorescence, absorption, diffraction, dispersion, and microscopy, the MIMS captures and measures the actual molecules present in the water. This advancement will enable the user to detect and measure the component parts that make up the oil, and provide detailed analysis of the water with respect to cocktails of oil and added chemicals in water, and nonfluorescing dissolved organic compounds simultaneously. A prototype of the instrument is shown in Figure 3.5.

Section III

Properties

4 Characterization of Athabasca and Arabian Light Vacuum Residues and Their Thermally Cracked Products

Implications of the Structural Information on Adsorption over Solid Surfaces

Francisco Lopez-Linares, Mazin M. Fathi,
Lante Carbognani Ortega, Azfar Hassan,
Pedro Pereira-Almao, Estrella Rogel, César Ovalles,
Ajit Pradhan, and John Zintsmaster

CONTENTS

CONTEXT

Changes in crude oil composition under cracking conditions can influence

- Thermal production processes
- Adsorption on porous solids
- Physical properties

ABSTRACT

The characterization of the thermal cracking products (visbreaking) from Athabasca vacuum residue (ATVR) and Arabian Light vacuum residue (ALVR) as a function of severity is presented. Information on elemental analysis (CHNSO), metal content (Ni, V), molecular mass (MM), aromaticity, total naphthenic carbon, total secondary methylene groups, chain length, and solubility parameter were determined to assess structural information. All of them were found to vary as a function of thermal cracking severity. For example, a reduction in H/C, MM, total secondary carbon, and chain length along conversion is observed. In addition, heteroatom contents also changed with residue conversion, leading to products enriched in nitrogen and oxygen but depleted in sulfur as severity increased. Interaction of such products with well-designed macroporous solid surfaces such as Ca–kaolin and Ca–Ba–kaolin indicated that adsorption uptakes depend on the structural properties of the products and on the textural properties of the solids. For both solids, adsorption uptakes for visbroken products from ATVR increase as a function of residue conversion. Parameters such as aromaticity, asphaltene content, and nitrogen content seem to be the drivers for adsorption uptake. In contrast, for ALVR, lower uptakes were found than in ATVR. For ALVR, adsorption also seems to be governed by aromaticity, asphaltene content, and nitrogen content of the visbroken products, as well as by surface properties.

4.1 INTRODUCTION

The surplus of conventional oils has decreased considerably in the last decade, motivating the pursuit for other alternatives such as bitumen, petroleum residues, and oil sands as alternative feeds for future upgrading processes. They are expected to be among the dominant fossil fuel energy alternatives during the next few decades [1]. Thus far, petroleum residues have been the least exploited feedstocks in conventional upgrading units owing to their intrinsic characteristics that require more severe process conditions. They are primarily constituted by high refractory compounds such as asphaltenes, which are claimed to be responsible for undesirable residua properties such as high viscosity. Additionally, high content of heteroatoms such as sulfur and nitrogen, as well as the presence of transition metals such as Ni and V on residua fractions, make processing more difficult.

Nevertheless, there are various well-established processes for converting heavy hydrocarbons into more profitable liquid and gas products. Some of them are based on thermal processes; delayed coking and visbreaking (VB; mild thermal cracking) have been extensively employed because of their maturity and ability to produce desired products at a good profit margin [2]. VB is characterized by low conversion yields since operational severity is limited by heater fouling and product instability, which have been associated with the asphaltene content on the feed [3]; on the other hand, delayed coking produces large percentages of coke that requires further management or disposal [4].

Nowadays, in the refining industry, hydroconversion and deasphalting processes are largely applied in upgrading residual oil [5]. Hydroconversion requires large hydrogen partial pressures to achieve profitable conversion levels with low coke production [6], while deasphalting processes require large amounts of precipitant solvents that could lead to process issues in terms of evaporation losses [7]. Owing to low refinery margins combined with the current crude oil prices, the oil community identified potential niches for this type of feedstocks, encouraging the search and deployment of new technologies for upgrading the bottom of the barrel.

To overcome the previously mentioned constraints, another feasible strategy proposed is the use of steam and tailored catalysts at given process conditions. This concept could economically promote higher conversion in thermal processes without the requirement for major equipment investments. For example, steam catalytic cracking (SCC) using unsupported ultradispersed catalysts has been reported as an alternative in upgrading vacuum residue (VR), with a higher conversion level than conventional processes, generating products with a quality good enough to be used in further refining processes [8]. This alternative has some particularly attractive advantages over conventional supported catalysts for heavy feedstock processing, by eliminating potential catalyst pore plugging and increasing the accessibility of highly dispersed active sites to large reactant molecules, which minimizes diffusion control and adds handling flexibility for the processed catalyst [9,10].

In a previous work, Athabasca VR (ATVR) and its thermally cracked products (VB) were characterized in terms of ^1H and ^{13}C nuclear magnetic resonance (NMR) average structural parameters, molecular weight, elemental analysis, and asphaltene solubility profile analysis [11,12]. Noticeable changes in asphaltene solubility profiles, aromaticity (fa), total naphthenic carbon (TNC), and secondary methylene groups depending on process severity were determined. This structural information was analyzed and correlated with adsorption on modified kaolin surfaces (Ca–kaolin and Ca–Ba–kaolin), as an initial step for any potential catalytic upgrading process. In the same context, preliminary characterization of Arabian Light VR (ALVR) and its products from thermal cracking, steam cracking, and SCC processing was recently reported [13].

The value of accessing structural and molecular information of feeds and products was demonstrated by Boduszynski et al. [14,15], as a tool to elucidate variations in the chemical composition of heavy petroleums as a function of their atmospheric equivalent boiling points up to 1400°F and residue solubility. This approach was used to determine the compositional changes occurring during hydrocracking for vacuum gas oil derived from a desulfurized residue [15]. Detailed compositional data helped understand the complex chemistry occurring during such processes, and explained why vacuum gas oil (produced in a high-severity residue desulfurization process, which is enriched in polycyclic aromatic hydrocarbons) was more difficult to hydrocrack than other VGOs produced at lower severity [15].

This work aims to characterize, in more detail, the thermally cracked products from ATVR and ALVR as a function of the severity of conversion, to gather structural information. Then, correlations of such properties during the interaction with modified kaolin surfaces were carried out as a fundamental step for understanding further catalytic processes. The final outcome of this research is illustrating the importance of obtaining structural information as a predictive tool for subsequent upgrading steps of these products.

4.2 EXPERIMENTAL

4.2.1 Safety Consideration

Before and during the preparation of the feedstocks and products, proper personal protective equipment (glasses, gloves, laboratory coat) must be worn when handling all chemicals, solvents, feedstocks, and products. The Material Safety Data Sheets of the chemicals must be available and covered before use. Waste materials should be disposed responsibly.

4.2.2 Materials and Methods

Spectrophotometric-grade toluene; high-performance liquid chromatography–grade methylene chloride, n-heptane, methanol, and tetrahydrofuran; and NMR-grade chromium(III) acetylacetonate ($Cr(acac)_3$), 97%, deutero-chloroform ($CDCl_3$), 99.96 atom% D, and tetramethyl silane, from Sigma-Aldrich, were used as received. A VR from Athabasca bitumen was provided by Suncor (Canada); this material contains 26 wt.% distillable fractions (545°C–) as determined by high-temperature simulated distillation. About 0.7 wt.% methylene chloride–insoluble materials were determined to be present within this residue.

Mild thermally cracked products (VB) were prepared at different conversion levels according to a previous report [16]. As an example, VB of ATVR was carried out at 380°C using a glass reactor. Varying reaction times allowed collecting products with diverse conversion levels, up to 35 wt.%. Yields and conversion levels were determined by weighting the reactor contents before and after VB. The stability of the products was determined by titration with n-cetane at 100°C (P-value), with the appearance of precipitated asphaltenes detected by visible (Vis) light microscopy carried out at ambient temperature. A National model DC3-163 optical microscope provided with a camera system was used. Monitoring was generally performed with a magnifying power of 40×.

A VR from Arabian light crude was provided by Saudi Aramco (Saudi Arabia); it contains 18.3 wt.% distillable fractions (545°C–) as determined by high-temperature simulated distillation. Mild thermal cracking (VB) at different conversion levels was successfully achieved for this residue, as reported elsewhere [10]. For example, the VB experiments were performed in a pilot plant described in detail in Reference 16. The severity was achieved by varying the temperature from 390°C up to 430°C and liquid hourly space velocity spanning from 1.5 h^{-1} up to 5.0 h^{-1}. The percent weight conversion reported here is based on the conversion to lighter products of 540°C+ hydrocarbons that are found in the original feed. As an example, the calculation is made by first determining the weight of the 540°C+ hydrocarbon (HC) present on the feed and in the heavy products from the bottom of the hot separator, using Equation 4.1:

$$\text{Conversion} = \frac{(540°C+)\,HC \text{ in feed}\,(g) - (540°C+)\text{ in heavy product}\,(g)}{(540°C+)\,HC \text{ in feed}\,(g)} \quad (4.1)$$

Characterization of the feeds and thermally cracked products was performed at Chevron facilities. Description of the techniques and conditions are found elsewhere [11]. Parameters such as aromaticity (fa), total secondary CH_2 (TSC), chain length, and TNC were obtained by ^1H and ^{13}C NMR using a Bruker Avance 500 spectrometer provided with a 5-mm BBI probe. Each sample was mixed 1:1 (wt./wt.) with $CDCl_3$. Carbon, hydrogen, nitrogen (CHN) elemental analysis was carried out with a Carlo Erba model 1108 analyzer. Determination of S, Ni, and V was performed by using inductively coupled plasma atomic emission spectrometry (ICP–AES). Either a Thermo Scientific ICP–AES model IRIS Intrepid XDL II (Radial) or an iCap 6500 series was used. Oxygen content was determined by using the difference from the CHNS reported values. Asphaltene solubility parameters were determined following a previously reported method [12].

Molecular weights were determined with gel permeation chromatography using a Mixed E column. A blend of 90:10 methylene chloride/methanol was used as a carrier. The tests were run at room temperature using an HP Series 1100 chromatograph and an Alltech ELSD 2000 detector. Measurements were made using dichloromethane solutions of the analytes at a fixed concentration of 1000 ppm. A series of aromatic molecules, including dyes and phorphyrins, were used as standards to obtain a calibration curve. A similar approach has been used successfully before [17].

Details of experimental procedures, adsorbent synthesis, and characterization have been published elsewhere [11]. Kinetic experiments were carried out by continuously measuring the UV–Vis absorbance changes of the adsorbate from toluene solutions in contact with adsorbents. As an example, an adsorbent in extrudate form is contacted with toluene solutions of the VR feedstock and VB product at a defined time. The instrument used for these experiments was a Cary 4E dual-beam spectrophotometer from Varian. Typically, the procedure was as follows: 3 mL of toluene solution of the corresponding adsorbate (60–70 ppm) was placed in a screw-cap cuvette (Spectrosil 3.5 m, rectangular cell quartz, open-top cap, light path: 10 mm, from Sigma-Aldrich) and the initial absorbance was measured (A0); then, 0.3 g of the solid was added, the system was closed, and the cuvette cell was placed in the spectrophotometer, and every 15 min a spectrum was recorded for a total period of 120 min in static mode at room temperature.

Kinetic plots were obtained by transforming the absorbance A(t) into relative absorbance RA(t) (to make it independent of the initial concentration) as function of time. RA(t) was transformed at solution concentration Cs(t) (mg/L), using Equation 4.2:

$$\text{Cs}(t) = \text{RA}(t)C_0 \quad (4.2)$$

where RA_0 and C_0 are the initial relative absorbance and concentration, respectively. The amount VR and VB adsorbed at any time Ca(t) per gram of adsorbent (mg/g) was calculated using Equation 4.3:

$$Ca(t) = C_0 - Cs(t)V/m \qquad (4.3)$$

where V is the solution volume (L) and m is the mass (g) of the adsorbent.

Typical errors in the absorbance scale do not exceed 10% relative. Details of the method have been published previously [18].

With the aim of simplifying the names for feed and products, the following notation will be used through the chapter:

Athabasca vacuum residue: ATVR
Arabian Light vacuum residue: ALVR
Athabasca thermally cracked product at a given conversion: ATVBx, where
    x = conversion
Arabian Light thermally cracked products at a given conversion: ALVBx,
    where x = conversion

4.3   RESULTS AND DISCUSSION

4.3.1   THERMAL CRACKING

The ability to access as much as possible detailed characterization of feedstocks, products, and their changes upon processing is desirable in any chemical process. For that reason, for several years, the Catalysis for Bitumen Upgrading group (University of Calgary) has been studying the upgrading of heavy oils and bitumen via thermal reactions assisted by steam and ultradispersed catalysts [8–10]. Recently, in a joint effort with Chevron, a detailed characterization of all products obtained in such reactions has been carried out.

As the outcome of this cooperation, a detailed structural information for Athabasca feedstocks and products have been determined, and the corresponding reactivity with adsorption on solid surfaces such as kaolins has been correlated [11]. To expand on this knowledge, the characterization of ALVR and their corresponding thermal products has been accomplished, and detailed information is presented in this chapter. Table 4.1 summarizes the corresponding properties determined for each feedstock.

From Table 4.1, differences among the residua are observed; that is, ATVR contains higher heteroatom (S, N, O), metal (Ni and V), and asphaltene content. Similarly, ATVR has higher molecular weight and naphthenic character than ALVR. On the other hand, the paraffinic nature of ALVR is reflected by its higher values of total secondary carbon and chain length. These properties are expected to influence the extent of residue conversion in which residues are simultaneously disproportioned into low MM distillates and hydrogen-deficient bottoms (visbroken residues) during VB conditions. VB yields were controlled by the stability of the corresponding visbroken products. As reported in the literature, the stability parameter was determined by optical microscopy observation of asphaltene precipitation using a well-known peptization parameter, i.e., P-value [19].

Having detailed information of the feedstocks before processing could help predict volumetric yields as well as the chemical and physical characteristics of the corresponding products. In Table 4.2, characterization results obtained for the VB

TABLE 4.1
Properties of Athabasca and Arabian Light Vacuum Residua

Parameter/Feedstock	Athabasca (ATVR)	Arabian Light (ALVR)
C (wt.%)	82.30	85.32
H (wt.%)	9.72	10.09
S (wt.%)	5.31	4.24
N (wt.%)	0.62	0.35
O (wt.%)	2.05	0.00
H/C	1.42	1.42
V (ppm)	201	82
Ni (ppm)	89	22
MM (Da)	1459	603
fa	0.33	0.28
Total secondary CH_2 (at 29.7 ppm)	19.50	26.18
Total naphthenic carbon (TNC)	8.84	4.26
Chain length (CL)	4.54	5.72
Solubility parameter ($MPa^{0.5}$)	17.00	17.23
Asphaltenes C7 (wt.%)	15.00	10.10

TABLE 4.2
Properties of Visbroken Products from ATVR

Parameter	Conversion (wt.%)				
	0.0	8.7	13.7	23.3	28.5
C (wt.%)	82.30	83.23	83.71	83.51	83.75
H (wt.%)	9.72	9.53	9.07	8.83	8.41
S (wt.%)	5.31	4.94	4.60	4.52	4.50
N (wt.%)	0.62	0.68	0.72	0.82	0.92
O (wt.%)	2.05	1.61	1.91	2.32	2.42
H/C	1.42	1.37	1.30	1.27	1.21
V (ppm)	201	298	296	326	347
Ni (ppm)	89	111	111	122	129
MM (Da)	1459	1107	830	693	689
fa	0.33	0.38	0.50	0.43	0.55
Total secondary CH_2 (at 29.7 ppm)	19.50	20.8	16.4	16.2	14.6
Total naphthenic carbon (TNC)	8.84	8.67	5.70	3.93	3.20
Chain length (CL)	4.54	3.88	3.96	4.52	3.79
Solubility parameter ($Mpa^{0.5}$)	17.0	17.3	17.7	17.8	18.2
P-value	2.6	1.7	1.5	1.2	1.1
Asphaltenes C7 (wt.%)	15	18	20	26	29

products from ATVR are presented. Under typical VB conditions, the upper residue conversion limits for this sample typically reach values of 30 wt.% and P-values close to 1.1, as reported with other feedstocks [19–21]. By comparison with ATVR feedstock (0% weight conversion), it can be seen that considerable changes have been promoted as a function of process severity.

Previous reports dealing with thermal conversion processes suggest that large molecules present in VR are cracked into smaller molecules, depending on the severity level [22–25]. Analyzing the results presented in Table 4.2, there is clear confirmation of the above-mentioned process, i.e., a reduction of the MM, chain lengths, and total secondary CH_2 groups, has occurred, indicating that alkyl appendages were cracked, leading to products with less alkyl substitutions [22]. The fact that aromaticity increased with the reduction of the TNC might be an indication of naphthenic ring aromatization by hydrogen production and/or H_2 transfer reactions. Additional indication of the quality of the products is given by stability as determined by solubility parameter profiling; higher values are found for the VB products, indicating a higher tendency to precipitate. These results are consistent with the lower P-values determined for these products, in comparison with those found for ATVR feed.

Table 4.3 presents the characterization results for the visbroken products from ALVR feedstock. It is observed therein that analogous to ATVR, thermal cracking promotes considerable changes on this feedstock as a function of residue conversion. Reductions in MM, increases in aromaticity (fa), and reduction in TNC, among others, are modified upon residue conversion.

Under the highest attained conversion (ALVR28.5), the product did not reach instability as determined by its P-value (1.2). This value is still above the agreed

TABLE 4.3
Properties of Visbroken Products from ALVR

Parameter	Conversion (wt.%)				
	0.0	20.3	21.7	23.6	28.5
C (wt.%)	85.32	84.53	84.80	84.75	84.09
H (wt.%)	10.09	9.90	9.94	9.84	9.61
S (wt.%)	4.24	4.61	4.03	4.41	4.59
N (wt.%)	0.35	0.36	0.36	0.38	0.37
O (wt.%)	0.00	0.60	0.86	0.63	1.34
H/C	1.42	1.41	1.41	1.39	1.37
V (ppm)	82	84	82	84	86
Ni (ppm)	22	28	23	24	23
MM (Da)	603	534	547	542	538
fa	0.28	0.39	0.41	0.36	0.35
Total secondary CH_2 (at 29.7 ppm)	26.18	24.48	25.79	28.35	28.15
Total naphthenic carbon (TNC)	4.26	3.03	4.14	3.21	2.71
Chain length (CL)	5.72	5.52	5.46	4.36	4.18
P-value	2.9	1.5	1.5	1.3	1.2
Asphaltenes C7 (wt.%)	10	12	13	14	15

instability limit (1.1). Analyzing other parameters such as aromaticity and naphthenicity, it seems that the products display more aromatic character as residue conversion increased. However, an interesting difference among the feeds is observed: the abundance of long (–(CH$_2$)–) moieties for ALVR did not change as a function of severity. This fact may have some implications for further upgrading steps. Additionally, the asphaltene contents varied less in absolute terms, which could lead to products with higher stability, as it is indeed reflected by the higher P-values for ALVR visbroken products in comparison with the original ALVR. However, a relative increase of asphaltene content was observed to be the same for visbroken ATVR and ALVR residua (~50% rel.) for products with 28.5% VB conversion.

Further comparison of selected parameters from Tables 4.2 and 4.3 allows identifying important differences among ATVR and ALVR virgin feedstocks and their visbroken products. In Figure 4.1, the MM in log scale as a function of VB severity are presented. It can be observed that both feeds suffered reduction in MM under mild cracking conditions. The same fact has been observed by other authors for asphaltenes treated under VB conditions [25]. It is interesting to note that ATVR suffered higher levels of MM reduction than ALVR. It is believed this is an indication of the higher reactivity of ATVR in comparison with ALVR. This behavior possibly affects the molecular structures that would have an impact later on the adsorption behavior. In fact, it has been observed that a reduction of ATVR and ALVR MM increases the adsorption uptake on macroporous kaolin surfaces [11,13].

Residua MM reduction has been associated with cracking alkyl appendages by temperature, via β-scission [22]. In this sense, total secondary methylene groups (TSC) were determined by ^{13}C NMR for both visbroken products and used as indicators of chain lengths. Figure 4.2 shows how this parameter varied as a function of VB severity. Initially, it is observed that ATVR has lower TSC values, meaning that these products display short chains in their structures in comparison with ALVR. At a conversion close to 29 wt.%, it is noticeable that the TSC for ALVR VB product doubles the value of the ATVR VB counterpart. Therefore, ALVR products have a more paraffinic character, and ALVR may react differently under thermal cracking conditions.

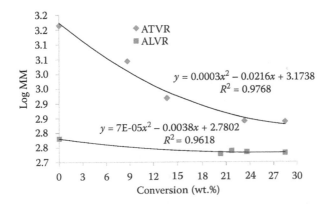

FIGURE 4.1 MM changes upon conversion during thermal cracking for ATVR and ALVR.

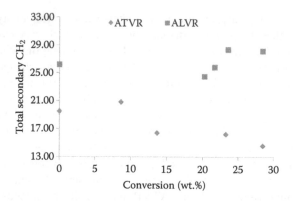

FIGURE 4.2 Behavior of total secondary CH_2 (TSC) during thermal cracking for ALVR and ALVR.

Furthermore, results show that the TSC for ATVR decreases progressively with increased thermal conversion. From the preceding discussion, the final product (ATVR28.5) has shorter alkyl appendages compared with the feed. Thus, it is expected that this product would present less steric hindrance during the interaction with solid surfaces. On the contrary, ALVR presents slightly higher TSC values as a function of thermal cracking severity, showing minor modifications of this parameter during the process. As mentioned, it is expected that this property would have some impact during adsorption—an aspect that will be covered in Section 4.3.2.

In a previous work with ATVR, it was found that the aromatic character of the products increased along with VB severities [11]. The gathered data for ALVR (Table 4.3) allowed studying if this feed displayed the same behavior. Figure 4.3 presents the aromaticity (fa) changes for both feedstocks as a function of VB severity, as determined by ^{13}C NMR.

The aromaticity of products from both residua increased as VB severity increase. However, ATVR leads to visbroken products with a higher aromatic character than

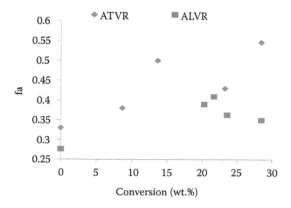

FIGURE 4.3 Changes in aromaticity upon conversion during thermal cracking for ATVR and ALVR.

TABLE 4.4

Degree of Aromaticity/Naphthenicity during Visbreaking for ATVB and ALVR

	Athabasca (ATVR)					Arabian Light (ALVR)				
	0	8.7	13.7	23.3	28.5	0	20.3	21.7	23.6	28.5
fa/TNC	0.04	0.04	0.09	0.11	0.17	0.06	0.13	0.10	0.11	0.13

that found for ALVR. In fact, at the same conversion level (28.5 wt.%), it can be seen that fa increased ~66% relative to ATVR in comparison with the ~25% relative to ALVR. The increase in aromaticity and lowering of TSC contents for ATVR suggest that ATVR visbroken products have exposed aromatic moieties and are able to have stronger interactions with solid surfaces and/or with other similar organic components.

^{13}C NMR analysis made it possible to determine the degree of naphthenicity and to correlate this parameter with the aromaticities. Thus, it was possible to obtain insights regarding the detailed molecular constitution in terms of aromatic/naphthenic characters. Table 4.4 shows that there is a crossover of this parameter for ATVR vs. ALVR fractions. Initially, ALVR shows a higher value, then displays a constant behavior that compares with intermediate conversions achieved for ATVR products. This last residue further proceeds toward increased aromaticities.

Considering that thermal cracking conversion is limited by asphaltene formation, in Figure 4.4 asphaltene content is plotted versus residue conversion for both feeds. Previous reports indicated that asphaltenes interact to a greater extent with solid surfaces [18,26–28]; thus, the knowledge of the amount of these refractory fractions is very important. As expected, the asphaltene content increases in both cases with a polynomial second-degree fitting. It is interesting that ATVR promoted a higher increase in asphaltene formation compared with ALVR with the concomitant generation of products with lower stability. Considering that we are looking to further study the interaction of these products with solid surfaces, it is possible to obtain

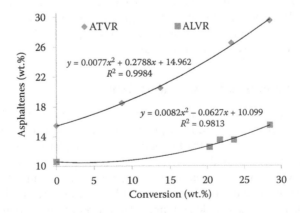

$$y = 0.0077x^2 + 0.2788x + 14.962$$
$$R^2 = 0.9984$$

$$y = 0.0082x^2 - 0.0627x + 10.099$$
$$R^2 = 0.9813$$

FIGURE 4.4 Asphaltene formation during thermal cracking for ALVR and ALVR.

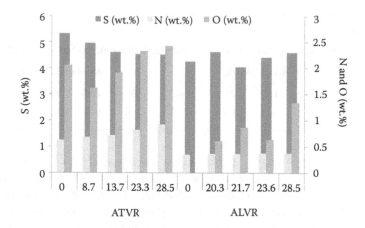

FIGURE 4.5 Variation of the heteroatom content with visbreaking severity for ATVR and ALVR.

preliminary information about the potential adsorption uptakes of these products. ATVB products should interact in greater extension than ALVB products since these materials have higher asphaltene contents, higher aromaticities, and fewer steric protecting groups, i.e., long –(CH$_2$)–, compared with ALVR products.

Figure 4.5 presents how the heteroatom contents, i.e., sulfur (left Y-axis), nitrogen, and oxygen (right Y-axis), varies on both feeds along the conversion.

Oxygen and nitrogen contents were found to increase as a function of thermal cracking severity in both feeds. Connecting these findings with those previously discussed, it seems that more severely visbroken products are mainly aromatic nuclei with heteroatoms substitutions within their structures. The presence of exposed heteroatoms probably facilitates the interaction with solid sorbent surfaces. In fact, it has been observed that increases in nitrogen contents in VB products induce higher adsorptive capacity over kaolin surfaces [11,18].

Additionally, it is observed that the sulfur content decreased on conversion in all cases studied. It is expected that during thermal processes, easily crackable sulfur moieties, like those present within sulfidic groups, are eliminated, which in turn leads to the production of light distillates plus heavier species. The latter will remain within the heavy bottoms of visbroken residua.

4.3.2 Adsorption over Macroporous Kaolins

Our research group has focused on the adsorption of different visbroken Athabasca feedstocks and their corresponding asphaltenes over different solid surfaces and minerals [11,13,16,18,29,30]. In the previous studies, emphasis was focused on the textural properties of the solids, such as surface area, pore size, and pore volume. Less attention was paid to the molecular characteristics of the adsorbates. The need for determining the structural information of adsorbates in greater detail was identified as a tool for a better understanding of the overall adsorption process, thus enhancing the feasibility of novel heavy oil upgrading schemes. Having gathered these information from ATVR and ALVR visbroken products, it was decided to study how these fractions interact with solid surfaces that could act as support or catalysts for steam gasification reaction [29]. Correlation between adsorption uptakes of ATVR with kaolin surfaces was determined, and it was found that higher uptakes were observed for adsorbates with higher aromaticity, solubility parameter, and nitrogen and oxygen contents. Trends between these molecular properties with uptakes allowed for the prediction of potential adsorption capacity of the thermally cracked products over this type of surfaces [11].

As an extension of the last studies, adsorption experiments of ATVR and ALVR feeds and their corresponding visbroken products were performed over two macroporous kaolins: Ca–kaolin and Ca–barium (Ba)–kaolin. These macroporous solids contain calcium and/or a mixture of calcium–Ba oxides within their structures. The main properties that could be relevant to adsorption are as follows: surface area (N_2-BET), 12 m^2/g (Ca–kaolin) and 18 m^2/g (Ca–Ba–kaolin); pore size, >100 nm; total acidity determined by NH_3–TPD, 520 mmol NH_3/g (Ca–kaolin) and 286 mmol NH_3/g (Ca–Ba–kaolin). The fact that these macroporous solids have relatively large pore diameters is an important aspect, since restrictions of large molecular weight molecules through the solid pore networks are not expected.

Table 4.5 presents the corresponding adsorption uptakes for each product in toluene solution having initial concentrations in the 60–70 mg/L range. Experiments were carried out at 295 K and monitored via UV–Vis, as described in Section 4.2. Reported uptakes were determined after 120 min of interaction. Typical errors in the absorbance scale do not exceed 10%. In addition, the main structural informations were included again in Tables 4.2 and 4.3 for a better visualization of their contributions on the adsorption uptakes.

From Table 4.5, it is observed that under the experimental conditions employed in this work, ATVB products have more adsorptive capacity than the ALVR counterparts. Regarding ALVR products, all of these materials are less adsorptive than the corresponding ATVRs and, on the other hand, these materials are more adsorptive for Ca–kaolin than Ca–Ba–kaolin. These findings could be a combination of the properties of the adsorbates and the surfaces studied. Potential explanations are addressed in the ensuing paragraphs. As mentioned in Section 4.3.1, the discussed information gave us indications of these behaviors. Parameters such as MM,

TABLE 4.5
Adsorption Uptakes for ATVR and ALVR and Their Visbroken Products over Ca–kaolin and Ca–Ba–kaolin

Parameter/	Athabasca					Arabian Light			
Conversion (wt.%)	0	8.7	13.7	23.3	28.5	0	20.3	23.6	28.5
Ca–kaolin[a]	0.117	0.119	0.127	0.132	0.139	0.084	0.093	0.116	0.125
Ca–Ba–kaolin[a]	0.143	0.149	0.16	0.169	0.206	0.079	0.069	0.071	0.063

[a] Update at 120 min (mg/g).

aromaticity, asphaltene and heteroatom contents, and total secondary CH_2 have been found to play important roles in the adsorption [11]. Figure 4.6 presents the effect of MM reduction on adsorption uptakes for the ALVR and its visbroken products.

For VB products from both residua, it is observed that the reduction of MM under mild cracking conditions leads to materials with higher capacity to interact with the solid surfaces studied. It is known that aromatic molecules tend to enhance the interaction via π-bonding through the aromatic electrons to silicon atoms present on the kaolin surface [31–33].

Regarding the interaction of ALVB products with Ca–Ba–kaolin, it seems that MM does not contribute strongly to the adsorption. It seems that other parameters related to the surface are controlling the adsorption for this family of products.

Therefore, size reduction can positively affect the uptakes by exposing aromatic moieties to the proximity of the available adsorption active site. Aromaticity is a good

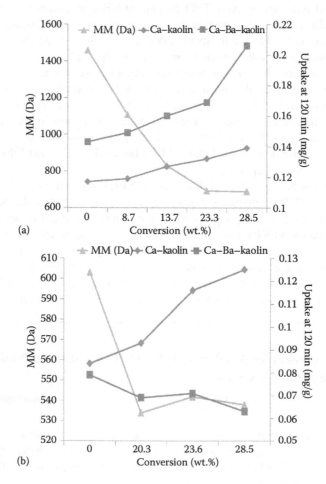

(a)

(b)

FIGURE 4.6   Adsorption uptakes for ATVR (a) and ALVR (b) and their visbroken products over Ca–kaolin and Ca–Ba–kaolin as function of the molecular mass (MM).

descriptor for aromatic moieties, being then correlated with uptakes, as presented in Figure 4.7.

From Figure 4.7, it is clear that ATVB products, which are enriched in aromatics, heteroatoms, and asphaltenes, interact more extensively with both solids. For ALVB products, it seems that the same effect prevails for Ca–kaolin but not for Ca–Ba–kaolin. The ALVB20.5 product has more aromaticity than the feed and the other products. However, ALVR products are not enriched considerably with heteroatoms (this will be discussed in the following paragraphs). One further fact is that Ca–Ba–kaolin is more basic in nature owing to the presence of the binary alkali earth oxide mixture. From all the preceding findings, it seems that aromaticity is not an important driver for the adsorption of ALVR visbroken products as it was for ATVR fractions.

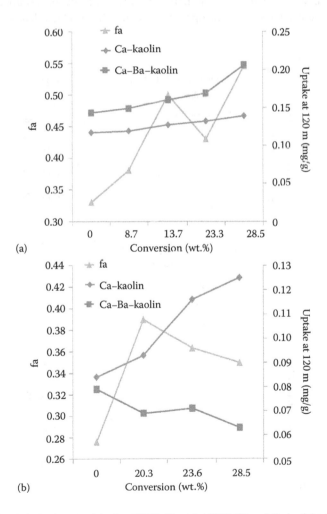

FIGURE 4.7 Adsorption uptakes for ATVR (a) and ALVR (b) and their visbroken products over Ca–kaolin and Ca–Ba–kaolin as a function of aromaticity (fa).

Considering that asphaltene fractions contain most of the heteroatoms present and show the highest aromaticities, it was decided to investigate if the asphaltene contents of the VB products would be a driver for the adsorption. Figure 4.8 presents the results for both residues.

As can be seen, the increase of asphaltene contents enhances the adsorption for most visbroken products. For ATVB the effect can be observed on both surfaces, and for the case of ALVB only on Ca–kaolin. It seems that for a given surface (e.g., Ca–kaolin), the adsorbate properties lead to products more prone to interact with this surface. Considering that both solids have different acidities (Ca–kaolin: 520 mmol NH_3/g, Ca–Ba–kaolin: 286 mmol NH_3/g), the correlation between adsorption uptakes and nitrogen contents was investigated, and the results are presented in Figure 4.9.

As shown, the correlations found between nitrogen contents and uptakes suggest that incremental increases in nitrogen content for the products leads to increased

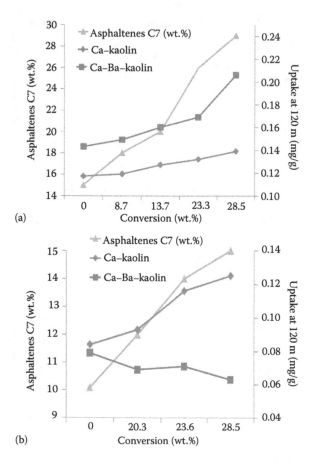

FIGURE 4.8 Adsorption uptakes for ATVR (a) and ALVR (b) and their visbroken products over Ca–kaolin and Ca–Ba–kaolin as a function of asphaltene content.

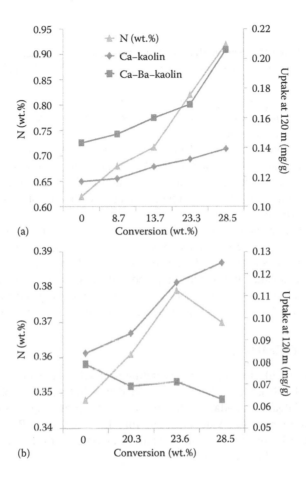

FIGURE 4.9 Adsorption uptakes for ATVR (a) and ALVR (b) and their visbroken products over Ca–kaolin and Ca–Ba–kaolin as a function of nitrogen content.

uptakes. This indicates that nitrogen compounds could effectively interact with the available acid sites on the surfaces. The fact that the total acidity is reduced in the Ca–Ba–kaolin may point out that a wider range of functional groups present on the ATVB products, such as aromatic amines and pyridine benzologs, among others, could interact effectively with this less acidic surface. ATVB28.5 is compared with ALVB28.5, and the ratio among their properties was determined in terms of the heteroatom contents, fa, and asphaltene content (S: 0.98, N: 2.49, O: 1.81, fa: 1.56, Asp: 1.73). It can be concluded that ATVB visbroken products will be more reactive toward kaolin-based surfaces owing to increased effects from these promoter parameters. At present, it is not clear which is (are) the reason(s) for the low adsorptivity determined for ALVB products over Ca–Ba–kaolin. This aspect that demands further investigation.

4.4 CONCLUSIONS

The results indicate that knowledge on structural properties of heavy oil adsorbates could help predict their potential adsorptive capacity over solids such as macroporous Ca–kaolin and Ca–Ba–kaolin.

Thermal cracking was determined to yield products with higher aromaticity, lower contents of naphthenic carbon, lower alkyl lengths, lower sulfur, and higher nitrogen and oxygen contents.

Findings with barium-modified macroporous kaolin (Ca–Ba–kaolin) indicated that this is a more adsorptive material for ATVB products, giving the same adsorption trends previously determined with the calcium analog. The results suggest that the studied structural parameters can be extended to different solids as feasible predictors for adsorption processes.

The adsorption uptakes for visbroken products from ATVR increase in both solids as a function of VB conversion. Parameters such as aromaticity, asphaltene content, and nitrogen content appear to be drivers for the observed adsorption. On the contrary, for ALVR, lower uptakes were found compared with ATVR. However, the same driver parameters, i.e., high aromaticity, asphaltene content, and nitrogen content of visbroken products, as well as surface properties, were identified to enhance adsorption.

ACKNOWLEDGMENTS

The authors would like to acknowledge funding from Alberta Ingenuity Centre for In Situ Energy (AICISE), and the facilities provided by Schulich School of Engineering, University of Calgary, Canada, and Saudi Aramco for providing oil samples. M.M. Fathi wishes to thank Saudi Aramco for funding this research. E. Rogel, C. Ovalles, A. Pradhan, and J. Zintsmaster wish to thank Chevron ETC for funding and providing permission to publish this work.

REFERENCES

1. Roberts, P. 2005. *The End of Oil: On the Edge of a Perilous New World*. Wilmington, MA: Mariner Books.
2. Speight, J. G., Özüm, B. 2002. *Petroleum Refining Processes*. New York: Marcel Dekker.
3. Higuerey, I., Rogel, E., Pereira, P. 2001. Residue stability study in a thermal catalytic steam cracking process through theoretical estimation of the solubility parameter. *Pet. Sci. Technol.* 19: 387–401.
4. Pereira, P., Marzin, R., Zacarias, L., Lopez, I., Hernandez, F., Cordova, J., Szeoke, J., Flores, C., Duque, J., Solari, B. 1998. Aquaconversion: A new option for residue conversion and heavy oil upgrading. *Vis. Tecnol.* 6: 5–14.
5. Mohanty, S., Kunzru, D., Saraf, D. N. 1990. Hydrocracking: A review. *Fuel* 69: 1467–1473.
6. Rana, M. S., Sámano, V., Ancheyta, J., Diaz, J. A. I. 2007. A review of recent advances on process technologies for upgrading of heavy oils and residua. *Fuel* 86: 1216–1231.
7. Le Page, J., Chatila, S. G., Davidson, M. 1990. *Resid and Heavy Oil Processing*. Paris: Editions Technip.
8. Pereira, P., Marzin, R., Zacarias, L., Cordova, J., Carrazza, J., Marino, M. 1999. U.S. Patent 5885441.

9. Pereira, P., Hill, J., Wang, J., Vasquez, A. 2005. *Ultra Dispersed Catalyst for Processing Heavy Hydrocarbon Fractions*. Atlanta, GA: AICHE, Spring National Meeting.
10. Fathi, M. M., Pereira-Almao, P. 2011. Catalytic aquaprocessing of Arab light vacuum residue via short space time. *Energy Fuels* 25: 4867–4877.
11. Lopez-Linares, F., Carbognani, L., Hassan, A., Pereira-Almao, P., Rogel, E., Ovalles, C., Pradhan, A., Zintsmaster, J. 2011. Adsorption of Athabasca vacuum residue and their thermal cracked products over macroporous solids: Influence of the molecular characteristics. *Energy Fuels* 25: 4049–4054.
12. Rogel, E., Ovalles, C., Moir, M. 2010. Asphaltenes stability in crude oils and petroleum materials by solubility profile analysis. *Energy Fuels* 24: 4369–4374.
13. Fathi, M. M., Lopez-Linares, F., Carbognani, L., Pereira-Almao, P., Rogel, E., Ovalles, C., Pradhan, A., Zintsmaster, J. 2013. Molecular properties of virgin and converted Arabian light vacuum residue: Implications on adsorption and catalysis. *Prep. Am. Chem. Soc. Div. Energy Fuels* 58: 512–515.
14. Boduszynski, M. M. 1998. Composition of heavy petroleums. 2. Molecular characterization. *Energy Fuels* 2: 597–613.
15. Sullivan, R. F., Boduszynski, M. M., Fetzer, J. C. 1989. Molecular transformations in hydrotreating and hydrocracking. *Energy Fuels* 3: 603–612.
16. Carbognani, L., González, M. F., Lopez-Linares, F., Sosa-Stull, C., Pereira Almao, P. 2008. Selective adsorption of thermal cracked heavy molecules. *Energy Fuels* 22: 1739–1746.
17. Guzmán, A., Bueno, A., Carbognani, L. 2009. Molecular weight determination of asphaltenes from Colombian crudes by size exclusion chromatography (SEC) and vapor pressure osmometry (VPO). *Pet. Sci. Technol.* 27: 801–816.
18. González, M. F., Sosa-Stull, C., López-Linares, F., Pereira-Almao, P. 2007. Comparing asphaltenes adsorption with model heavy molecules over macroporous solid surfaces. *Energy Fuels* 21: 234–241.
19. diCarlo, S., Janis, B. 1992. Composition and visbreakability of petroleum residues. *Chem. Eng. Sci.* 47: 2695–2700.
20. Omole, O., Olieh, M. N., Osinovo, T. 1999. Thermal visbreaking of heavy oil from the Nigerian tar sand. *Fuel* 78: 1489–1496.
21. Singh, J., Kumar, M. M., Saxena, A. K., Kumar, S. 2004. Studies on thermal cracking behavior of residual feedstocks in a batch reactor. *Chem. Eng. Sci.* 59: 4505–4515.
22. Wiehe, I. A. 1994. The pendant-core building block model of petroleum residua. *Energy Fuels* 8: 536–544.
23. Thomas, M., Fixari, B., LePerchec, P., Princic, Y., Lena, L. 1989. Visbreaking of Safaniya vacuum residue in the presence of additive. *Fuel* 68: 318–322.
24. Zhang, L., Yang, G., Que, G., Zhang, Q., Yang, P. 2006. Colloidal stability variation of petroleum residue during thermal reaction. *Energy Fuels* 20: 2008–2012.
25. Dettman, H., Inman, A., Salmon, S., Scott, K., Fuhr, B. 2005. Chemical characterization of GPC fractions of Athabasca bitumen asphaltenes isolated before and after thermal treatment. *Energy Fuels* 19: 1399–1404.
26. Acevedo, S., Ranaudo, M. A., Escobar, G., Gutierrez, L., Ortega, P. 1995. Adsorption of asphaltenes and resins on organic and inorganic substrates and their correlation with precipitation problems in production well tubing. *Fuel* 74: 595–598.
27. Acevedo, S., Ranaudo, M. A., Garcia, C., Castillo, J., Fernandez, A. 2003. Adsorption of asphaltenes at the toluene–silica interface: A kinetic study. *Energy Fuels* 17: 257–261.
28. Pernyeszi, T., Dekany, I. 2001. Sorption and elution of asphaltenes from porous silica surfaces. *Colloids Surf. A Physicochem. Eng. Asp* 194: 25–39.
29. Sosa, C., Gonzalez, M. F., Carbognani, L., Perez-Zurita, M. J., Lopez-Linares, F., Moore, R. G., Husein, M., Pereira, P. 2006. Visbreaking based integrated process for bitumen upgrading and hydrogen production. Paper 2006-074. CIPC 2006 (Canada International Petroleum Conference).

30. Lopez-Linares, F., Carbognani, L., Sosa-Stull, C., Pereira-Almao, P., Spencer, R. J. 2009. Adsorption studies in Athabasca core sample: Virgin and mild thermal cracked residua. *Energy Fuels* 23: 3657–3662.
31. Wolkow, R. A., Moffat, D. J. 1995. The frustrated motion of benzene on the surface of Si(111). *J. Chem. Phys.* 103: 10696–10700.
32. Carbone, M., Piancastelli, M. N., Zanoni, R., Comtet, G., Dujardin, G., Hellner, L. 1998. A low symmetry adsorption state of benzene on Si(111)7×7 studied by photoemission and photodesorption. *Surf. Sci.* 407: 275–281.
33. Mirji, S. A., Halligudi, S. B., Sawant, D., Patil, K. R., Gaikwad, A. B., Pradhan, S. D. 2006. Adsorption of toluene on Si(1 0 0)/SiO$_2$ substrate and mesoporous SBA-15. *Colloids Surf A Physicochem. Eng. Asp* 272: 220–226.

5 Analysis of Olefins in Heavy Oil, Bitumen, and Their Upgraded Products

*Lante Carbognani Ortega, Francisco Lopez-Linares,
Qiao Wu, Marianna Trujillo, Josune Carbognani,
and Pedro Pereira-Almao*

CONTENTS

CONTEXT

- The presence of olefins can be a "show stopper" during the transportation of crude oils
- Olefins can polymerize, creating gums and solid deposits
- Olefin issues can be found throughout the petroleum value chain

ABSTRACT

Bitumen and vacuum residual upgrading improves transportability, process-ability, and revenue. Thermal or catalytic upgrading gives rise to olefins that can polymerize, creating solid deposits. Olefin determination for heteroatom-enriched, wide-distilling bitumen-upgraded samples is addressed here. Two methodologies were selected: bromine number titration (Br#) and olefinic-H determination via ^{1}H-NMR spectroscopy. Br# analysis was found slightly affected by aromatic hydrocarbons and strongly dependent on heteroatomic species. The estimation of mono-olefins via ^{1}H-NMR spectroscopy was found feasible, as universal calibration proposed herein.

Conjugated diolefins were observed to affect sample stability, being their estimation attempted via ^{1}H-NMR. However, aromatic hydrogen signals over-lap the range of conjugated diolefinic-H in the ^{1}H-NMR spectrum, demanding extra efforts for successful analysis.

Selected samples illustrate the applicability of the developed ^{1}H-NMR methodologies, showing advantages from catalytic partial upgrading versus pure thermal processing. Analytical aspects need addressing and optimization; further efforts for unraveling olefin/diolefin levels and variables that trigger solid deposition in upgraded bitumens are required.

5.1 INTRODUCTION

Alkenes (olefins) are reactive compounds commonly absent from petroleum crude oils, except in a few cases when reservoirs have been exposed to magma intrusions and/or radiolysis [1,2]. Small-molecular-weight (MW) olefins are produced during oil-refining schemes, which imply thermal processing [3]. Downhole production of olefinic compounds also occurs as a consequence of production mechanisms such as shale oil production via retorting [4] and in situ combustion applied to oil reservoirs [5]. Alkenes, and particularly conjugated alkenes, are unwanted components since they easily polymerize, giving rise to gum formation and solid deposition [6].

Depletion of conventional oil and increased production of heavy and extra-heavy oils (the latter is identified as "bitumen" in Canada) have led to partial upgrading of these materials for their transportation to large refining facilities where all the upgrading processes are installed [7]. Partial upgrading in field facilities can pro-duce olefins, and currently, a limit of 1 wt.% is established as a specification for petroleum producers [8].

Determination of olefins and diolefins by chemical methods was common decades ago [9–11]. High-performance liquid chromatography [12–15] and supercritical fluid chromatography [16,17] became popular during the 1980s. Spectroscopy techniques such as near infrared [18] and nuclear magnetic resonance (NMR) [19–21] have been described for the characterization of alkenes in petroleum fractions. Group contribu-tion methods have been proposed to estimate the NMR chemical shifts of olefinic protons [22]. The determination of olefins in petroleum products has been the subject of several reviews [23–25].

Methods for olefin determination have been standardized [26,27], with bromine number (Br#) determination being the most applied [27]. Br# determination was initially applied to distillates boiling at <315°C; however, modifications have been described to make it suitable for fractions boiling up to 550°C [27,28]. The influence of structural effects and heteroatoms on Br# analysis of small-MW compounds have been reported [27,29]. Correlations of Br# with the percentage abundance of olefins have been published, with values from 1/1.8 to 1/2.9 times Br# being described to provide olefin contents in wt.% and/or vol.% [30–32]. For complex materials such as whole bitumen for which structural and heteroatomic effects cannot be ascertained, a standard ¹H-NMR method is practiced by petroleum producers [8]. This method requires two spectra per sample, the first one for the neat material and the second for the 1-decene spiked sample. The method reports mass% olefins as 1-decene equivalents [8].

Upgrading heavy oil distillates, vacuum residua, and whole bitumens is currently studied in our laboratories to increase oil production, improve transportation, and increase oil revenues [33–39]. Partial upgrading for achieving transportability viscosity limits (350 cSt at a defined temperature) provides products that span wide distillation ranges (from naphthas to vacuum residua), and contain noticeable amounts of heteroatomic species [34–39]. The objective of this chapter is to present an analysis of olefinic compounds in partially processed heavy oil and bitumen fractions [40]. Br# determination [27] and ¹H-NMR will be studied, with ¹H-NMR ultimately being selected for the estimation of mono-olefin and conjugated diolefin abundances. Variables affecting ¹H-NMR determinations are discussed in greater detail, and selected examples are presented to illustrate the applicability of the developed methods.

5.2 EXPERIMENTAL

5.2.1 SAMPLES AND STANDARD OLEFINS

Athabasca bitumen and its derived fractions were analyzed. The studied samples were derived from different upgrading processes [33–39] and will be directly described in Section 5.3.

5.2.2 STANDARD AROMATIC AND HETEROCYCLIC COMPOUNDS

Standard hydrocarbons were provided by Sigma-Aldrich, Fluka, or Alfa Aesar, and used as received (i.e., without any further treatment). The following is a list of the studied compounds, with their source in parentheses and identification numbers in square brackets, as given in Figures 5.1 through 5.3.

The studied aromatics included the following: benzene (Sigma-Aldrich, 270709) [#1], toluene (Sigma-Aldrich, 34866) [#2], o-xylene (EMD, XX020-5) [#3], 1-Me-naphthalene (Fluka, 67880) [#4], fluorene (Aldrich, 12,833-3) [#5], phenanthrene (Aldrich, R11409) [#6], pyrene (Aldrich, 185515) [#7], chrysene (Fluka, 35754) [#8], benzo[a]pyrene (Sigma, B1760) [#9], benzo[k]fluoranthene (Aldrich, 12488) [#10], naphto(1,2,3,4-DEF)chrysene (Aldrich, R308803) [#11], dibenzoDEF-p-chrysene (Aldrich, R308668) [#12], and coronene (Aldrich, C84801) [#13].

The studied heteroatomic compounds were grouped into the following families:

- Sulfur compounds: *sec*-butylsulfide (Aldrich, 12,431-1) [#1], *n*-butylsulfide (Aldrich, B10,179-6) [#2], and dibenzothiophene (Aldrich, D32202) [#3]
- Oxygen compound: 2,3-benzofuran (Aldrich, B8002) [#4]
- Neutral N compound: carbazole (Sigma-Aldrich, C5132) [#5]
- Aromatic amines: 4-Et-aniline (Aldrich, E12001) [#6], 1-naphthylamine (Sigma, N9005) [#7], 1-amino-anthracene (Aldrich, A38606) [#8], and 2,3-diaminonaphthalene (Aldrich, 13,653) [#9]
- Basic N: benzoquinoline (Alfa Aesar, A17794) [#10], 5,6,7,8-tetrahydroiso-quinoline (Aldrich, 301981) [#11], 3-aminoquinoline (Aldrich, 232289) [#12], and 2-aminopyridine (Aldrich, A77997) [#13]
- S, O heterocyclic: diphenylsulfone (Fluka, 43330) [#14]
- O, N heterocyclic: 6-MeO-quinoline (Aldrich, 183067) [#15]
- S, N heterocyclics: 4-methylthioaniline (Aldrich, M54503) [#16] and 2-quinolinethiol (Aldrich, 116270) [#17]
- N, S, O heterocyclic: quinoline-65 (Aldrich, 641766) [#18]

5.2.3 KNOWN HYDROCARBON MIXTURES

Four hydrocarbon mixtures were prepared for calibration purposes, as described in the following:

- Paraffins mixture
- Aromatics mixture
- Wide olefins mixture
- Light olefins mixture

A description of these mixtures is presented in Table 5.1. These hydrocarbon mixtures were used for preparation of blends with known olefin contents whose Br# and mono-olefinic-H via ^{1}H-NMR were determined. The blends will be described in detail in Section 5.3.1. The following are the sources for the standard hydrocarbons not already presented in Section 5.2.2:

n-hexane (Sigma-Aldrich, 293253), *n*-heptane (Sigma-Aldrich, 650536), *i*-octane (Sigma-Aldrich, 258776), (*cis*+*trans*)-decaline (Sigma-Aldrich, 294772), Me-cyclohexane (Sigma-Aldrich, M37889), *n*-dodecane (Sigma, D221104), *n*-tetradecane (Aldrich, 17,245-6), *n*-hexadecane (Sigma-Aldrich, H6703), 2-3-diMe-butene-2 (Aldrich, 129259), cyclohexene (Sigma-Aldrich, 29240), α-Me-styrene (Aldrich, M80903), isoprene (Aldrich, I19551), 1-decene (Aldrich, D1807), 1-octadecene (Aldrich, 74740), *p*-xylene (Sigma-Aldrich, 134449), *n*-propylbenzene (Aldrich, 82119), tetraline (Sigma-Aldrich, 429325), and biphenyl (Fischer, O-1421).

In addition, a conjugated diolefin blend was prepared containing the following compounds (the number in parenthesis before the compound name indicates the wt.% of each component in the blend):

(13.96%) Isoprene (Sigma-Aldrich I19551, 99% purity), (23.70%) 2,4-diMe-1,3-pentadiene (Sigma-Aldrich,126551, 98% purity), (25.43%) 1,3-cyclohexadiene

TABLE 5.1

Prepared Hydrocarbon Mixtures

Paraffins		Wide Olefins		Aromatics		Light Olefins	
Compound	Wt.%	Compound	Wt.%	Compound	Wt.%	Compound	Wt.%
n-Hexane	7.45	2,3DiMe-butene-2	17.09	Benzene	5.32	2,3DiMe-butene-2	21.70
n-Heptane	32.57	Cyclohexene	22.36	Toluene	25.76	1-Hexene	58.99
i-Octane	22.96	αMe-styrene	10.36	p-Xylene	52.79	Cyclohexene	19.31
n-Octane	12.75	Isoprene[a]	9.82	n-Propylbenzene	6.34		
(c+t)Decaline	7.91	1-Decene	37.46	Tetraline	2.31		
Me-cyclohexane	7.97	1-Octadecene	2.91	Biphenyl	1.90		
n-Dodecane	3.71			1-MeNaphthalene	5.57		
n-Tetradecane	2.16						
n-Hexadecane	2.54						

[a] Isoprene's Hs on conjugated carbons were used for delimitation of diolefin/mono-olefin regions in the spectra.

(Sigma-Aldrich C100005, 97% purity), (11.65%) 1,2,3,4-tetraMe-1,3-cyclopenta-diene (Sigma-Aldrich 424471, 85% purity), and (25.25%) 1,3-hexadiene (Sigma-Aldrich 527408, 95% purity).

5.2.4 Bromine Number Determination

Br# was determined following a standard method [27]. One model T70 Mettler Toledo titrator was used for this purpose. The setup temperature for the analysis spanned the 0–5°C range (water–glycol recirculating bath).

5.2.5 NMR

^{1}H-NMR analyses were carried out at 298 K either with a Bruker Avance 400-MHz spectrometer (olefin content >2 wt.%) or with a Bruker Avance III 600-MHz spectrometer (olefin content <2 wt.%). The sample solvent was C_6D_6 (Sigma-Aldrich 175870, 99.96 atom%D), using 450 µL solvent for a 150 mg sample. Spectra were recorded using a 30° flip angle, applying pulses at 5-s intervals with 200 scans co-added for each spectrum. The spectrum reference peak is at 7.16 ppm (exchanged-normal benzene protons). Internal standard 1-decene, when used, was of ≥99.5% purity (Fluka P/N 30649). The wt.% of mono-olefinic-H was calculated using integrated areas from the ^{1}H-NMR spectrum, by using the expression

$$\text{Wt.\% mono-olefinic-H} = (\text{area } 4.5–6.3 \text{ ppm}) \times 100/(\text{total spectrum integrated area})$$
$$(5.1)$$

Conjugated diolefin analysis was exclusively carried out with the Bruker Avance III (600-MHz) system. Selected conjugated diolefinic proton signals appearing in the 6.3–6.8 ppm zone (protruding from the aromatic hump) were manually integrated and compared with dioxane-added internal standard (IS). The protocol for diolefin content determination is described in Section 5.3.3. Dioxane (IS) for conjugated diolefin estimation (Sigma-Aldrich 266309, 99.8% purity) was prepared in C_6D_6 (20% vol./vol.), with 5.0 µL added on top of the 160.0 mg sample for each run.

Two-dimensional (2D) H-1/X proton-carbon NMR correlation was carried out using the Brucker hsqcetgp program, ran with the Bruker Avance III (600-MHz) spectrometer.

5.2.6 Simulated Distillation

High-temperature simulated distillation by gas chromatography was performed according to published conditions [39]. The present experiments were started at −20°C to improve light-ends discrimination.

5.2.7 Hydrocarbon Group-Type Separation by Solid-Phase Extraction (SPE)

Olefin concentrates from light distillation cuts were isolated at the micropreparative scale following published solid-phase extraction (SPE) procedures [32]. Saturates +

olefins were eluted from SiO_2 cartridges with 1.5 dead volumes of *n*-pentane. Collected *n*-pentane solutions were directly percolated through a second cartridge packed with $SiO_2/AgNO_3$ 85:15 wt./wt.; saturates were eluted with additional 1.5 dead volumes of *n*-pentane (*n*-C5), and olefins were recovered with 2 dead volumes of methylene chloride (DCM). Analytical-grade solvents (*n*-C5 Sigma-Aldrich cat. no. 34,956, >99%; DCM Sigma-Aldrich cat. no. 320269, >99.5%) and silica (40 Å, 35–70 mesh; Aldrich cat. no. 24,217-9) were used. Adsorbents were activated before use by heating for 4 h inside an oven maintained at 100°C. Isolated sample solutions were directly tested via simulated distillation (see Section 5.2.6) to determine the carbon number spans for the isolated fractions.

5.3 RESULTS AND DISCUSSION

The discussion will be presented in five parts: (i) review of the Br# test for olefin determination of upgraded bitumen and bitumen fractions, which led to the use of ^{1}H-NMR due to the determined interferences from aromatics and, particularly, the presence of heteroatomic species; (ii) development of an ^{1}H-NMR mono-olefin analysis protocol and assessment of the factors affecting the method; (iii) development of an ^{1}H-NMR conjugated diolefin analysis protocol; (iv) illustration of the application of the developed ^{1}H-NMR methods for monitoring bitumen upgrading; and (v) brief review of the status of analytical methods for olefin analysis in upgraded bitumen, highlighting future actions to overcome the identified barriers.

5.3.1 BR# OLEFIN DETERMINATION AND INFLUENCE OF AROMATIC AND HETEROCYCLIC SPECIES

Olefin content determination with the widely used Br# test [27] seems convenient at first because the technique and the instruments involved are simple. However, validation of the method was undertaken to check an automatic titrator acquired in our laboratory. To this end, three sets of experiments were carried out to test the validity for (i) pure standard olefins, (ii) pure standard aromatics, and (iii) pure standard heterocyclic hydrocarbons.

Figure 5.1a presents the determined Br# for standard olefinic compounds. A biunivocal correlation was observed, indicating that the technique is sound and the equipment is responding as presumed. Deviation for the included conjugated diolefin (isoprene) was observed, with such deviations being already reported with the standard method [27]. The precision (Figure 5.1b) was found reasonable for most cases, except for samples having a low average Br#; for these latter cases, the percent relative standard deviation (%RSD) was found to span up to 6%. Detection of low levels of olefinic compounds was not found to be a problem, as suggested by the results plotted in Figure 5.1c.

Standard aromatic hydrocarbons were analyzed for Br#, and the results are presented in Figure 5.2. It was observed that most of the tested compounds provided Br# falling within the 2 ± 2 range; that is, larger than the expected zero response. One hypothesis formulated to explain these results was that the aromatic rings were brominated to certain levels under the setup conditions; to test this hypothesis,

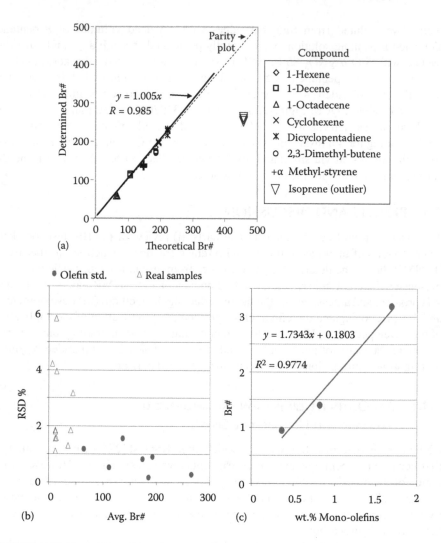

FIGURE 5.1 Bromine number determination: (a) analysis of standard olefins; (b) determined repeatability for triplicated analysis of standard olefins and real samples; (c) analysis of low-olefin-content samples (wide mono-olefin mixture [see Table 5.1] diluted in benzene).

the reaction was carried out under varying temperatures. Results determined for toluene and α-Me-naphthalene within the range −7°C to +15°C failed to show any differences. At present, no explanation exists for the contribution of small aromatics to Br# values; a suggestion was made to consider that the zero value can span the 0–4 range from this cause. One of the compounds showed an unexpectedly high Br# (benzo[a]pyrene, #9 in Figure 5.2). Two different batches were tested, giving Br# of 59 and 63. The stoichiometry indicates that one unidentified double bond within its fused aromatic core is acting as an olefinic bond, incorporating

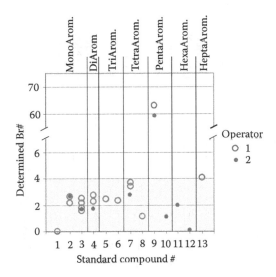

FIGURE 5.2 Bromine numbers determined for standard aromatic compounds. Key numbers identify each compound presented in Section 5.2.2.

two bromine atoms during the titration. In fact, it has been reported that carcinogenic compounds such as benzo[a]pyrene seem to undergo a two-stage metabolic activation [41], initially involving the formation of the 7,8-dihydrodiol [42] through a double bond, which is then further oxidized to give 7,8-dihydro-7,8-dihydrobenzo[a]pyrene-9,10-oxide [43].

There are indications for the standard method [27] about the response of heterocyclic hydrocarbons to Br# determination. Several heteroatomic hydrocarbons were analyzed to appraise this possibility. Figure 5.3 presents the results achieved. More than half of the evaluated compounds showed dramatic responses, showing Br# between about 100 and 300. The stoichiometry for two compounds was calculated to obtain a rough indication of how many Br atoms were incorporated during the titration: for carbazole (compound #5 in Figure 5.3), about four Br atoms were found incorporated; for 2,3-diamino-naphthalene (compound #9 in Figure 5.3), about seven Br atoms were incorporated. Considering that N, S, and O heteroatoms are abundant in heavy oil distillates and are major components of bitumen and oil residua [38,44], estimation of the contributions from these species to the determined Br# was attempted.

Regarding heavy oil distillates such as vacuum gas oil, typical aromatic contents span the 25–80 wt.% range. Polar compounds (or "resins") typically can reach up to 10 wt.%. Considering the Br# determined above for single compounds (Figures 5.2 and 5.3), the contribution of Br# to heavy distillates can be estimated, as presented in Figure 5.4. Up to about 35 units on the Br# scale can be added on top of the determined value for olefin content of a heavy oil fraction. In the ensuing discussion, analysis of real samples will add further evidence to the positive deviation of Br# determination that can be ascribed to the presence of heteroatoms.

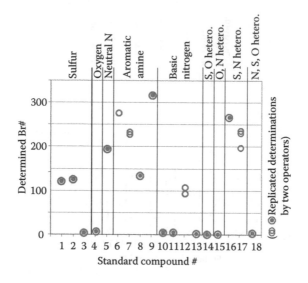

FIGURE 5.3 Bromine numbers determined for standard heteroatomic compounds. Key numbers identify each compound presented in Section 5.2.2.

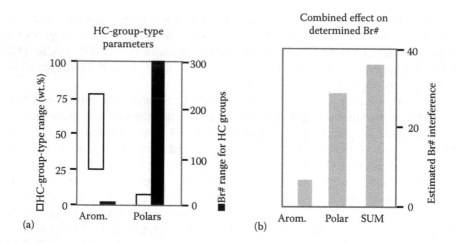

FIGURE 5.4 Estimated influence of aromatic and heterocyclic compounds on Br# determination. (a) Mass abundance and found Br# spanned by hydrocarbon groups. (b) Combined effect from group-type mass abundance times Br# spanned range. Vacuum gas oil is the material considered in the example.

Br# was determined for known hydrocarbon blends prepared with the mixtures presented in Table 5.1. Also, the percentage of mono-olefinic-H for these mixtures was correlated to the known mono-olefin abundances, with both correlations presented in Figure 5.5. Linear correlations were determined in both cases, as presented in the plots of Figure 5.5. Combining those linear correlations, it was possible to determine one expression correlating Br# versus mono-olefinic-H contents:

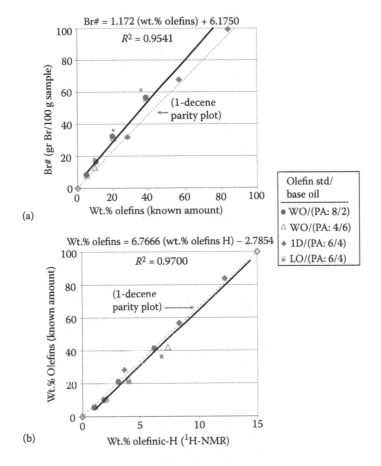

(a)

(b)

FIGURE 5.5 Br# (a) and mono-olefinic-H (b) determined for known hydrocarbon blends. Mixtures presented in Table 5.1 were used for blend preparation. The parity plot for commonly used 1-decene internal standard was appended over the plots.

$$Br\# = 7.9308 \, (\text{wt.\% mono-olefinic-H}) + 2.9106 \qquad (5.2)$$

Equation 5.2 was then used to understand the behavior of real samples containing abundant heteroatomic S and N species. Figure 5.6a shows that all the analyzed samples gave positive deviations of Br# versus the expected values based on the determined mono-olefinic-H content. Figure 5.6b and c show the same plot, with S and N elemental contents appended, respectively. Polar S and N functionalities are herein deemed as responsible for the positive deviations, since the Br# versus mono-olefinic-H correlation (Equation 5.2) was determined for pure hydrocarbons, i.e., only C- and H-containing compounds. In addition, two findings from these plots were deemed unexpected: (i) light materials and wide distillates spanning the C5–C60 range deviate, which is not expected in routine Br# determinations [27,28], and (ii) heavy residual fractions most enriched in S and N were not those that deviate to a large extent. At present, it can be speculated that particular polar functionalities

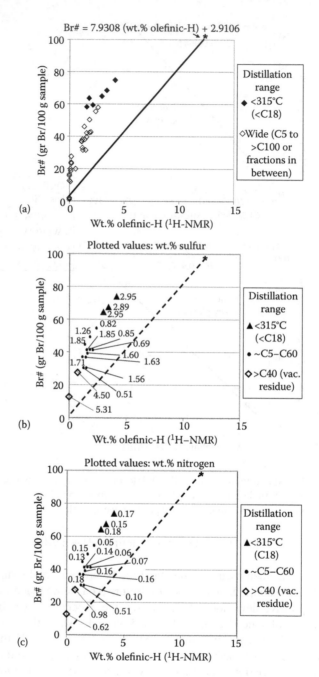

FIGURE 5.6 Br# and wt.% mono-olefinic-H determined for selected real samples. (a) Determined values for the two parameters. (b, c) Plots of (a) where S and N elemental contents were respectively appended. Parity plot between Br# and wt.% mono-olefinic-H is presented in all panels.

of lighter cuts are responsible for the larger observed deviations, while those present within the residua are not as important.

The findings discussed within this section indicate that contributions from aromatics, and particularly heteroatomic species, are important and overdetermined the Br#. This led to consider Br# as a nonconvenient tool for monitoring olefins in partially upgraded bitumen and bitumen-derived fractions. From the preceding discussion, another tool already described in the open literature was evaluated, i.e., ^{1}H-NMR [19–21].

5.3.2 ^{1}H-NMR for Mono-Olefin Analysis of Standard Mixtures and Upgraded Bitumen Fractions

The first aspect that must be addressed for the analysis of multicomponent mixtures like those studied here is how wide (or multidispersed) they are in molecular terms. This aspect was addressed by analyzing light fractions (340°C–) from upgraded bitumens. Olefins are better detected in these light fractions, since normally they are small-MW compounds that partition into lighter distillation cuts. Figure 5.7 presents gas chromatograms determined for two such cuts and their SPE-separated fractions. Olefin traces are observed to cover a wide MW range spanning from C5 up to C18 (no. of carbon atoms per molecule). These findings suggest that compounds spanning this range should be selected for calibration purposes.

In addition to the above-mentioned findings, a theoretical influence of MW and isomerism over the abundance of mono-olefinic-H determined via ^{1}H-NMR was

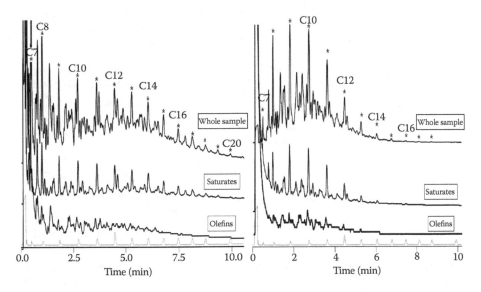

FIGURE 5.7 Carbon numbers spanned by olefinic compounds in light cuts (340°C–) isolated from upgraded bitumens. Analysis was carried out by gas chromatography simulated distillation. Asterisks highlight *n*-paraffin peaks. Hydrocarbons groups were separated via SPE.

carried out. Standard olefins mixed with polydispersed mixtures of standard aromatics and paraffinic compounds were studied. Calculations were made using six different base oils, different olefins, and three olefin concentration ranges (10, 25, and 40 wt.%). Base oils covering different distillation ranges, and different ratios of paraffins/aromatics were included. Figure 5.8 presents the outcome of these theoretical estimations. The following aspects were deduced from the estimations:

1. Small-MW mono-olefinic compounds increased, to a larger extent, the content of mono-olefinic-H because the relative contribution of the double bond within smaller molecules is larger.
2. Terminal olefins contribute more to mono-olefinic-H content because their contribution to olefinic-H is 3 compared with 2 for cyclic/branched olefins.
3. Light saturates and aromatic hydrocarbons present in blends provided lower mono-olefinic-H values because the total hydrogen contributed by these compounds is higher.
4. Higher saturate contents provided slightly lower mono-olefinic-H values because the contribution of H from paraffins is higher compared with the contribution from aromatics.

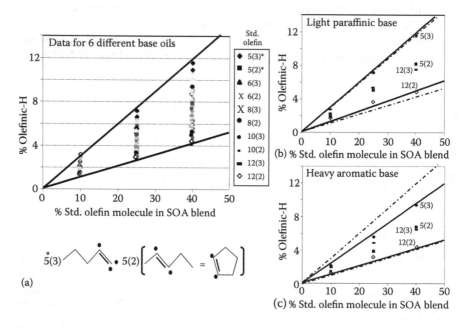

FIGURE 5.8 Theoretical estimation of mono-olefinic-H abundance in hydrocarbon mixtures. Standard olefins selected for blending (described in Section 5.3.2), identified with the notation illustrated for C5 compounds (bottom of [a]). SOA: saturates, olefins, aromatics. Appended solid lines in (a) highlight the limits determined for mono-olefinic-H contents for all studied base oils. Examples presented on (b) and (c) illustrate the limits for the selected olefin compounds in different base oils (dotted lines in these panels correspond to the whole limits presented in [a]).

5. For the same percentage of mono-olefin contents, a dramatic variation in mono-olefinic-H was determined, spanning from 1% to 3% mono-olefinic-H at the 10% concentration level and between 4% and 12% mono-olefinic-H at a 40% concentration level.

The above findings, particularly that presented in point 5, suggest that for complex hydrocarbon mixtures spanning wide MW ranges (mono-olefins of varying MW and varying terminal/internal double-bond distributions), a robust single universal mono-olefin calibration is difficult to achieve. These findings suggested preparing olefin calibration mixtures including compounds with varying isomerism spanning the C5–C18 carbon range, as determined for some of the real samples.

Table 5.1 presents four hydrocarbon mixtures used for preparing blends with known wt.% of mono-olefin contents for further analysis via ^{1}H-NMR. Tri- and tetra-substituted mono-olefins were not considered since their concentrations are generally very low. Diolefins (conjugated diolefins) will be considered in Section 5.3.3. Another series of mixtures containing wide olefins mixture were prepared in a blend of 80:20 (wt./wt.) saturates/aromatics; another series of mixtures with the wide olefins mixture were prepared in a blend of 40:60 (wt./wt.) saturates/aromatics. These blends aimed to unravel the possible interferences from aromatics, as previously discussed by others [31]. One series of mixtures contained known amounts of pure 1-decene in a base of saturates/aromatics 60:40 (wt./wt). 1-Decene is the selected compound for one proposed standard addition method [8], and is an average compound for the C5–C18 olefins spanned range (see Figure 5.7). The fourth prepared series contained known amounts of light olefins mixture in the same base of saturates/aromatics 60:40 (wt./wt.). Although these light olefins were not detected via gas chromatography simulated distillation, as discussed in preceding paragraphs (see also Figure 5.7), these light olefin blends were included considering that thermal cracking processes in general provide an abundance of low-MW alkenes (<C6). In addition to the preceding considerations, different base oils were evaluated since the ratio of paraffins/aromatics and the hydrogen content of these blends [45] have been reported to affect olefin determinations. Results determined with a 400-MHz spectrometer for high-olefin samples (2–85 wt.%) and a 600-MHz spectrometer for those with low olefin content (0.3–2.2 wt.%) are presented in Figure 5.9.

The results presented in Figure 5.9 indicate that similar linear calibration trends are observed for low or high olefin contents, independently of the base oil and olefin mixture used for calibration purposes. Figure 5.10 further presents statistical information that confirms linearity for the whole combined range of olefin contents; that is, the measured versus predicted values agreed (Figure 5.10a) and residuals were found to be evenly distributed (Figure 5.10b). However, the calculated %RSD (Figure 5.10c) indicates that at >10 wt.% olefin contents, the errors are <10%; for lower contents, precision is a matter of concern. A "universal" calibration is proposed from the above findings (Equation 5.3). High olefin contents are deemed reliably reported with this method; however, its poor precision, particularly for low olefin contents, can only allow considering it as an estimation rather than a robust analytical method.

$$\text{Wt.\% mono-olefins} = 6.5008 \, [\% \text{ mono-olefinic-H}] - 0.1947 \qquad (5.3)$$

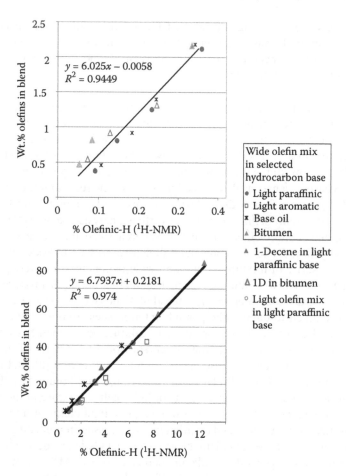

FIGURE 5.9 Mono-olefin contents determined via ¹H-NMR. Standard compounds selected for calibration blends presented in Table 5.1. High olefin contents were analyzed with a 400-MHz spectrometer; low olefin contents were reported with a 600-MHz spectrometer. 1D, 1-decene; Olefin mix, wide and light olefin mixtures included in Table 5.1.

As mentioned above, a 600-MHz spectrometer was chosen for low olefin contents (<2 wt.%) because it allows better signal separation and also, more important, because it provides better signal-to-noise ratio (S/N; about four times higher). This allowed further understanding of how the hydrogen elemental content of the samples affects the analysis, as reported before [45]. Figure 5.11 shows that, indeed, hydrogen contents affect the outcome of ¹H-NMR determination of mono-olefinic-H; each base oil with different hydrogen content provides a different linear correlation (Figure 5.11a). Calibration slopes were found to increase with increasing H content (Figure 5.11b). These findings again indicate that the universal calibration discussed before (Equation 5.3) is a good average response for mono-olefin content estimation of widely diverse samples; however, this is not a definitive answer for this scope.

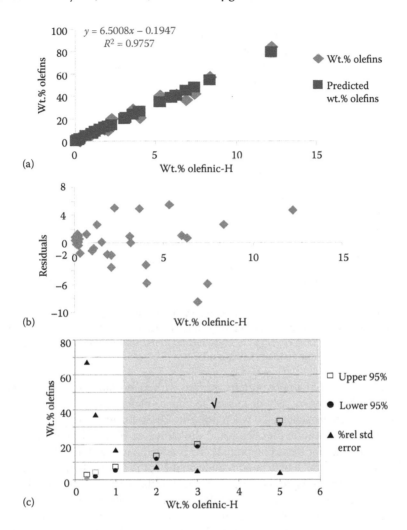

FIGURE 5.10 Statistical analysis for % mono-olefinic-H determined via ^{1}H-NMR. Data for low olefin contents (600-MHz spectrometer) combined with data for high olefin contents (400-MHz spectrometer). (a) Wt.% olefinic-H line fit plot, (b) wt.% olefinic-H residual plot, and (c) statistical analysis.

5.3.3 CONJUGATED DIOLEFIN ANALYSIS WITH ^{1}H-NMR

As mentioned before, conjugated diolefins reportedly are the most deleterious olefins present in thermally treated oil materials [6]. Regions where mono-olefinic [1,19,20] and conjugated diolefinic [46] protons appear in the ^{1}H-NMR spectrum have been reported, indicating that the latter substantially overlap the large aromatic proton hump that spans from about 6.6 to 9.0 ppm. Figure 5.12a presents the ^{1}H-NMR spectrum for a mixture of diolefins (compounds in the mixture described in Section 5.2.3). Only the protons linked to the internal conjugated carbons are displaced to the lower field; that is, they appear in the range from 6.3 to 6.8 ppm referred in the

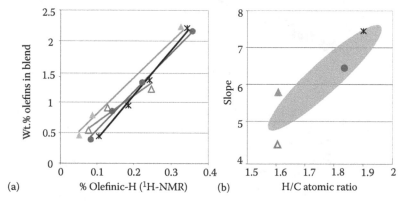

$\triangle$ 1D in bitumen $y = 4.357x + 0.236$ $R^2 = 0.9635$

$\blacktriangle$ Olef mix bitumen $y = 5.925x + 0.2085$ $R^2 = 0.995$

$\bullet$ Olef mix/paraf base $y = 6.4762x - 0.1755$ $R^2 = 0.9961$

$\ast$ Olef mix/base oil $y = 7.5094x - 0.3961$ $R^2 = 0.9975$

(a)   % Olefinic-H (^{1}H-NMR)   (b)   H/C atomic ratio

FIGURE 5.11 Slope of calibration curves for mono-olefinic-H as a function of elemental hydrogen content of the samples: (a) wt.% Olefinic-H determined for known mixtures prepared in different base oils; (b) slope of correlations found for known olefinic mixtures as a function of the H/C atomic ratio of the base oil.

literature [46]. The remainder protons overlap with mono-olefinic signals, meaning that the correlation previously discussed for mono-olefins (Section 5.3.2, Equation 5.3) also considers these signals, when present. These features further allow understanding why Br# titration for conjugated diolefins (see Figure 5.1a) provides about one-half of the titer for these compounds: Br is added onto the 1,4 carbons of conjugated diolefins and one double bond remain in the 2,3 positions, being inactivated to further Br reaction owing to the inductive effect of the already added two Br atoms. Figure 5.12b and c illustrate how complex diolefin signals can be further qualitatively discriminated with correlated proton–carbon 2D spectra for a better selection of the conjugated diolefinic signals used for analysis. In all cases, the diolefinic protons attached to internal conjugated carbons are displayed within the 6.3–6.8 ppm range in the ^{1}H-NMR domain.

The development of a calibration curve for diolefin content determination proved to be a difficult task because integration of diolefin "spike signals" over the aromatic hump was unreliable using the software provided with the spectrometer. The problem was solved by using the decades-old manual integration method. Upgraded bitumen with known amounts of conjugated diolefin mixtures were prepared and analyzed via ^{1}H-NMR. The averaged sum of the heights of several selected diolefinic signals was compared to a constant spike of dioxane (IS) and correlated with the known conjugated diolefin abundances. Figure 5.13a presents the analysis protocol and Figure 5.13b plots the found linear correlation. At least eight conjugated diolefin signals are commonly taken for analysis; the correlated height ratio versus diolefin abundance is presented in Equation 5.4. Equation 5.5 is the linear correlation herein proposed as a way to estimate conjugated diolefin contents in upgraded bitumen or

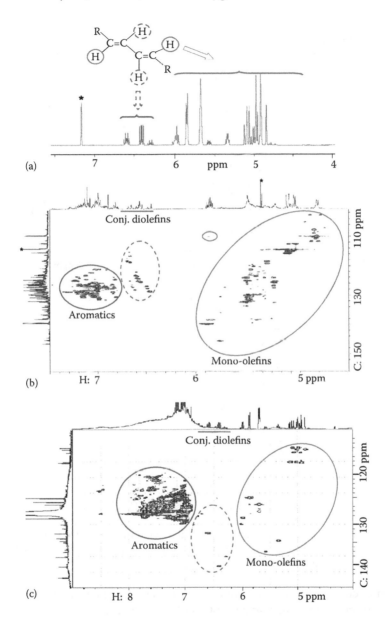

FIGURE 5.12 Conjugated diolefin analysis via ^{1}H-NMR (600-MHz spectrometer). (a) Mixture of standard conjugated diolefins (see Section 5.2.3). (b) Proton–carbon NMR correlation for light cut (340°C–) isolated from vacuum residue thermal cracking. (c) Proton–carbon NMR correlation for upgraded bitumen with spiked (1.93 wt.%) standard conjugated diolefin mixture. Asterisk identifies solvent peak (when detectable).

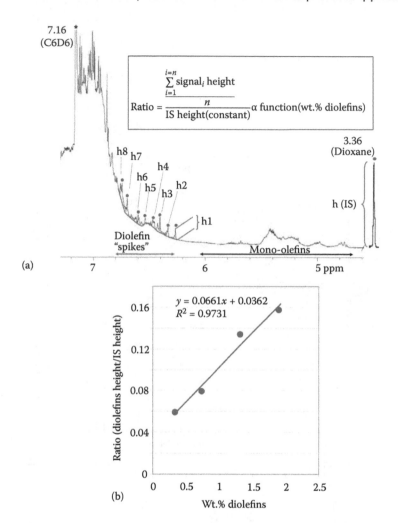

FIGURE 5.13 (a) Conjugated diolefin signals integration protocol. (b) Linear correlation between determined conjugated diolefin height ratio versus dioxane (IS) and known conjugated diolefin concentrations.

bitumen fractions. The line does not pass through the origin, meaning that the contribution from the aromatic hump is not totally eliminated.

$$\text{Ratio} = \frac{\dfrac{\sum\limits_{i=1}^{i=n} \text{signal}_i \text{ height}}{n}}{\text{IS height(constant)}} \, \alpha \text{ function(wt.\% diolefins)} \qquad (5.4)$$

$$\text{Ratio avg. conj. diolefin heights/IS height}$$
$$= 0.0661 \, (\text{wt.\% conjugated diolefins}) + 0.0362 \qquad (5.5)$$

5.3.4 Applications of ¹H-NMR to Olefin Monitoring in Petroleum Upgrading

Three examples are presented in this section to illustrate the application of the ¹H-NMR methods developed for upgraded bitumen monitoring. First, Figure 5.14 presents the analysis for the light cut (340°C–, Figure 5.14a) and the residue (340°C+, Figure 5.14b) from visbroken Athabasca vacuum residue. As expected, olefins selectively partition into the light end, which has a total olefin content of about 20 times larger compared with the heavier cut. Conjugated diolefins were observed in both fractions, with the determined amount in the light cut being noticeable. One aspect that confirms the difficulty of the interpretation of Br# is provided by the ¹H-NMR-determined total olefin contents and the respective Br#: the ratios of Br# to wt.% total olefins for the light and heavy cuts are 2.5 and 18, respectively, suggesting that the large amount of aromatic + polar material incorporated Br during the Br# titration of the heavy end.

A second example is presented in Figure 5.15, where synthetic partially upgraded bitumens containing thermally or catalytically cracked light materials are compared.

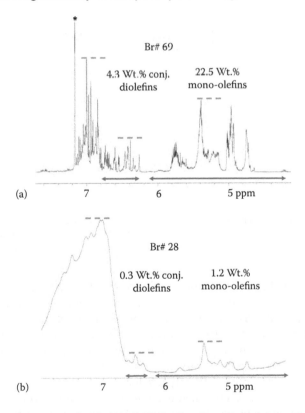

FIGURE 5.14 Olefin analysis via ¹H-NMR for fractions derived from thermal cracking of Athabasca vacuum residue: (a) light cut (340°C–); (b) residue (340°C+). Appended broken lines highlight height differences between aromatic/olefinic signals for both samples. Asterisk: solvent peak (C_6D_6).

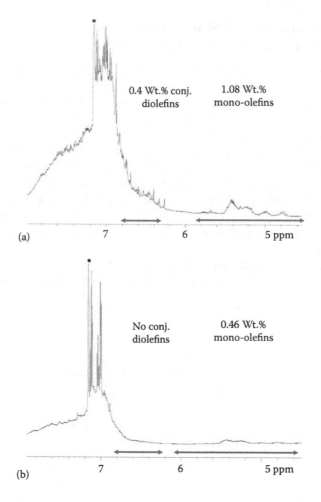

FIGURE 5.15   Olefin analysis via ¹H-NMR for synthetic bitumen containing (a) light fractions produced via thermal cracking and (b) light fractions produced via low severity–long residence time hydroprocesssing with NiMoW dispersed catalysts. Asterisk: solvent peak.

In the first case, higher levels of mono-olefins were determined and, more important, conjugated diolefins were observed. The upgraded bitumen that incorporated mild hydroprocessed light ends presented lower mono-olefin levels and no detectable conjugated diolefins. The latter finding is deemed important, showing that the use of dispersed active metals in low H_2 pressure process schematics, as described elsewhere [33,34], is able to provide products with expectably lower olefin polymerization potential during storage and transportation.

The final example here presented deals with the observed problems caused by conjugated diolefins. Figure 5.16 shows how solid deposition was observed after carrying out ¹H-NMR analysis for upgraded bitumen spiked with known amounts of conjugated diolefin mixtures. The sample containing 1.93 wt.% of conjugated diolefins produced severe solid deposition, with the solids being insoluble in a good solvent

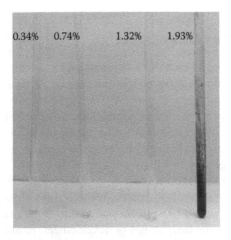

FIGURE 5.16 Images of ^{1}H-NMR tubes after analyzing samples containing from ~0.3 to 2.0 wt.% conjugated diolefins in upgraded bitumen. Conjugated diolefin mixture described in Section 5.2.3. After NMR analysis, tubes were cleaned with methylene chloride. Sample solutions were prepared and stored for 24 h inside a freezer (no light, 4°C). Time span between sample preparation and analysis completion: 30 h.

(methylene chloride). The appearance of these problems was randomly observed for conjugated diolefin levels of higher than about 1 wt.%; however, no systematic monitoring of the causes triggering the random appearance of solids has been carried out to date. The effects of olefin/conjugated diolefin on gasoline and gas–oil stability is a topic that has been extensively researched for many decades [47–50], with standard test methods described for such purposes [51–53]. However, for whole bitumen and partially upgraded bitumen, investigations are lacking, and immediate efforts are deemed necessary.

5.3.5 FINAL CONSIDERATIONS

From Section 5.3.4, a systematic understanding of olefin/conjugated diolefin concentration levels and the ultimate causes leading to solid deposition was forwarded as a requirement demanding immediate actions. Two other topics are here forwarded, which are believed to be worthy of attention of everyone interested in partially upgraded bitumen routinely pipelined and stored before entering final upgrading units.

The first topic is defining standard analytical protocols for mono-olefin and conjugated diolefin analysis that are agreed on by all interested parties in terms of bitumen upgrading, transportation, storage, and characterization. Research carried out before [54,55] and discussed in greater detail in Sections 5.3.1 through 5.3.4 showed that, thus far, estimation methods for these species are available; however, they are not deemed accurate and precise enough for specifying these species in commercial applications, implying low levels of these analytes. Detectability with modern NMR spectrometers is not an issue; that is, testing mono- and conjugated diolefin mixtures spiked at the 0.1 wt.% level over bitumen showed that mono-olefin signals are detectable without any problem (S/N with the 600-MHz spectrometer found at about 15). Conjugated diolefins, on the other hand, are more difficult to determine at low concentrations; in these cases, the S/N for the 0.1 wt.% spiked mixture (600-MHz spectrometer) was found to span the 2–4 range, and some of the peaks disappeared under the aromatic hump, suggesting that the conjugated diolefin ^{1}H-NMR limit of detection for complex mixtures is around 0.2 wt.%. A "universal calibration" agreed on by all interested parties should be found. One possible suggestion is to use one single standard compound (e.g., 1-decene) for mono-olefin analysis. The calibration correlation is thus simplified, being in this case the expression presented in Equation 5.6. One further suggestion to improve accuracy is to look in more detail at the effect of the hydrogen elemental content of the samples (see Figure 5.11); the use of a correction factor for taking this variable into account will appear necessary, with high probability. Conjugated diolefin determination in bitumen-upgraded samples is deemed to be still in the early stages of development, requiring further efforts to achieve a reliable analysis.

$$\text{Wt.\% mono-olefins} = 6.67 \ (\text{wt.\% mono-olefinic-H}) \qquad (5.6)$$

The second point raised pertains to a standard procedure that oil producers are currently using [8]. We found several points of concern regarding this method, here discussed and submitted to a greater audience for review. One concern is the use of the method for samples containing widely varying amounts of olefins. The method indicates the addition of 1 wt.% 1-decene, then comparing spectra with/without IS addition [8]. We found that for real samples containing greater than about 10 wt.% mono-olefins, the spectrum difference in areas is very small, incorrectly providing low mono-olefin contents for the tested materials (always <2 wt.%), as shown in Figure 5.17a. Another point of concern is the use of a single-point IS addition. Using the calculation formula of the standard method [8], different amounts of added IS were found to provide different contents of olefins, as shown in Figure 5.17b. By using standard addition, as described in analytical textbooks, several IS quantities were added and tested. Figure 5.17c shows that for concentrations around the actual one existing in the sample, the method is sound; however, larger IS amounts provide incorrect results because the abundance of aliphatic-H introduced by 1-decene overcomes the effect of the smaller mono-olefinic-H content provided by 1-decene (each added 1-decene molecule contributes 3 olefinic-Hs and 17 aliphatic-Hs). For this latter option, several ^{1}H-NMR spectra (about three) are required, thus making the approach less attractive for routine purposes.

The whole point raised in this final section of the chapter is that the determination of olefins in complex materials such as bitumen and upgraded bitumen still demands further efforts. Isolation of olefins that partition into lighter cuts will possibly be required for their improved detectability and optimized analysis. SPE followed by detailed analysis as suggested by others regarding light cuts [56] can be of help, despite such long and complex analytical schemes being a matter of concern for routine applications. ^{13}C-NMR is another possibility worth exploring despite the long acquisition times necessary; in ^{13}C-NMR, the higher signal resolution and better achievable S/N ratios, compared with ^{1}H-NMR, can probably be a trade-off for avoiding sample preseparation.

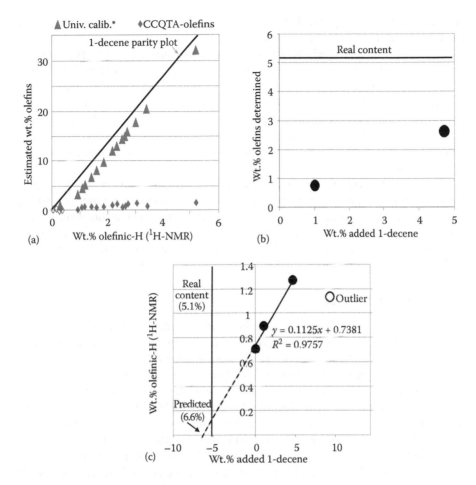

FIGURE 5.17 Estimated wt.% olefin calculated from mono-olefinic-H content determined via ¹H-NMR. (a) Calculation made via direct ¹H-NMR integration as proposed in this chapter ("universal calibration"; Equation 5.3) or via the proposed standard method (CCQTA-Olefins [8]). (b) CCQTA calculated olefin content with varying spiked amount of 1-decene. (c) Calculated olefin content using standard addition method (the same data points used in [b] were used in this example; only the calculation approach varied).

5.4 CONCLUSIONS

Olefin determination for heteroatom-enriched, wide-distilling bitumen and bitumen-upgraded samples was studied in this chapter. Br# determination was initially studied for olefin content determination; from the determined interferences caused by aromatic and particularly heteroatomic species on Br# titers, the technique was sidestepped to look for a more reliable method. ¹H-NMR spectroscopy was found feasible for mono-olefin analysis, being a universal calibration expression proposed in this chapter. Many parameters were found to affect ¹H-NMR determination of mono-olefinic-H contents; thus, the universal calibration was proposed only as a reasonable estimation, particularly for olefin-enriched samples (>10 wt.% olefins), for which relative errors were found to be <10%.

Conjugated diolefin analysis was also found feasible with ^{1}H-NMR; however, overlapping aromatic signals led to a manual integration protocol and the use of internal standard addition to be able to come up with a workable alternative.

Selected examples illustrated how the developed ^{1}H-NMR methods provide important understanding about the presence of olefins in thermally and/or catalytically upgraded bitumen samples. The upgrading schematics of low-severity dispersed catalysts were observed to provide synthetic bitumen with negligible conjugated diolefin levels.

Future work pertaining to olefin analysis in upgraded bitumen should address three important topics: (i) development of a highly accurate/precise and widely accepted calibration method(s) based on ^{1}H-NMR; (ii) finding the concentration levels of mono-olefin/conjugated diolefins and parameters triggering solid deposition in partially upgraded bitumen; and (iii) reviewing the currently used standards for olefin measurement in bitumen and partially upgraded bitumen.

ACKNOWLEDGMENTS

The authors would like to acknowledge the funding from the NSERC/Nexen/AIEES Industrial Research Chair in Catalysis for Bitumen Upgrading. The contribution of facilities from the Canada Foundation for Innovation; Institute for Sustainable Energy, Environment and Economy; Schulich School of Engineering; and Faculty of Science at the University of Calgary are also acknowledged. Colleagues from the Process Development team within the Catalysis for Bitumen Upgrading (CBU) group provided the samples studied in this chapter. Discussions held with Nestor Zerpa (Nexen Inc.), Phil Heaton and Michael Gordon (Maxxam, Calgary), and Francisco DaSilva (University of Calgary) are greatly appreciated. Dr. Michelle Forgeron (University of Calgary) is thanked for providing training in the use of NMR spectrometers. Dr. Hector Guzman (University of Calgary) is thanked for S and N elemental analysis. The kind invitation from Drs. Carl Rechsteiner and César Ovalles (Chevron ETC, Richmond, CA, USA) to participate in this project is greatly appreciated.

REFERENCES

1. Curiale, J. A.; Frolov, E. Occurrence and origin of olefins in crude oils: A review. *Org. Geochem.*, **1998**, 29(1–3), 397–408.
2. Frolov, E. B.; Smirnov, M. B. Unsaturated hydrocarbons in crude oils. *Org. Geochem.*, **1994**, 21(2), 189–208.
3. Speight, J. G. Thermal cracking, Chapter 14. In: *The Chemistry and Technology of Petroleum.*, 3rd edition. Marcel Dekker Inc.: New York, 1999; pp. 565–584.
4. McKay, J.; Latham, D. R. High-performance liquid chromatographic separation of olefin, saturate, and aromatic hydrocarbons in high boiling distillates and residues of oil shale. *Anal. Chem.*, **1980**, 52, 1618–1621.
5. Ryabov, V. D.; Dauda, S.; Kolesnikov, Y.; Karakhanov, R. A.; Rastova, N. V. Olefins from Karazhanbas crude produced by in situ combustion. *Chem. Technol. Fuels Oils*, **1996**, 32(2), 109–113.

6. Pereira, R. C. C.; Pasa, V. M. D. Effect of mono-olefins and diolefins on the stability of automotive gasoline. *Fuel*, **2006**, 85, 1860–1865.

7. Pereira-Almao, P.; Larter, S. An integrated approach to on site/in situ upgrading. In: *19th World Petroleum Congress.* Madrid, Spain, June 29–July 3, 2008.

8. CCQTA-Olefins in Crude Oil by Proton NMR Method (Canadian Crude Quality Technical Association, November 1, 2005), also identified as MAXXAM: CAPP Olefin by NMR version 1.04, November 2005 (Canadian Association of Petroleum Producers).

9. Suatoni, J. C. Determination of normal alpha-olefins by hydrobromination. *Anal. Chem.*, **1963**, 35(13), 2196–2198.

10. UOP Method 326–1965. Diene value by maleic anhydride addition reaction. **1965**, 1982 UOP Inc.

11. Poirier, M. A.; George, A. E. Method for determining the olefinic content of the saturate and aromatic fraction of petroleum distillates by hydroboration. *Fuel*, **1981**, 60(3), 194–196.

12. Alfredson, T. V. High performance liquid chromatographic column switching techniques for rapid hydrocarbon group-type separations. *J. Chromatogr.*, **1981**, 218, 715–728.

13. Miller, R. L.; Ettre, L. S. Quantitative analysis of hydrocarbons by structural group type in gasolines and distillates. *J. Chromatogr.*, **1983**, 259, 393–412.

14. Hayes, P. C. Jr.; Anderson, S. D. hydrocarbon group type analyzer system for the rapid determination of saturates, olefins and aromatics in hydrocarbon distillate products. *Anal. Chem.*, **1986**, 58, 2384–2388.

15. Hayes, P. C. Jr.; Anderson, S. D. Paraffins, olefins, naphthenes and aromatics analysis of selected hydrocarbon distillates using on-line column switching high-performance liquid chromatography with dielectric constant detection. *J. Chromatogr.*, **1988**, 437, 365–377.

16. Norris, T. A.; Rawdon, M. G. Determination of hydrocarbon types in petroleum liquids by supercritical fluid chromatography with flame ionization detection. *Anal. Chem.*, **1984**, 56, 1767–1769.

17. Albuquerque, F. C. Determination of conjugated dienes in petroleum products by supercritical fluid chromatography and ultraviolet detection. *J. Sep. Sci.*, **2003**, 26, 1403–1406.

18. Lopez-Garcia, C.; Biguerd, H.; Marchal-George, N.; Schildknecht-Szydlowski, N. Near infrared monitoring of low conjugated diolefins contents in hydrotreated FCC gasoline streams. *Oil Gas Sci. Technol., Rev. IFP*, **2007**, 62(1), 57–68.

19. Cookson, D. J.; Smith, B. E. Determination of the structure and abundances of alkanes and olefins in Fischer–Tropsch products using ^{13}C and [1]H n.m.r. methods. *Fuel*, **1989**, 68(6), 776–779.

20. Altbatch, M.; Fitzpatrick, C. P. Identification of conjugated diolefins in fossil fuel liquids by [1]H n.m.r. two-dimensional correlated spectroscopy. *Fuel*, **1994**, 73(2), 223–228.

21. Sarpal, A. S.; Kapur, G. S.; Mukherjee, S.; Tiwari, A. K. PONA analysis of cracked gasoline by ^{1}H NMR spectroscopy. Part II. *Fuel*, **2001**, 80, 521–528.

22. Matter, U. E.; Pascual, C.; Pretsch, E.; Pross, A.; Simon, W.; Sternhell, S. Estimation of the chemical shifts of olefinic protons using additive increments—II. The compilation of additive increments for 43 functional groups. *Tetrahedron*, **1969**, 25, 691–697.

23. Badoni, R. P.; Bhagat, S. D.; Joshi, G. C. Analysis of olefinic hydrocarbons in cracked petroleum stocks: A review. *Fuel*, **1992**, 71(3), 483–491.

24. Kaminski, M.; Kartanowicz, R.; Gilgenast, E.; Namiesnik, J. High-performance liquid chromatography in group-type separation and technical or process analytics of petroleum products. *Crit. Rev. Anal. Chem.*, **2005**, 35, 193–216.

25. deAndrade, D. F.; Fernandez, D. R.; Miranda, J. L. Methods for the determination of conjugated dienes in petroleum products: A review. *Fuel*, **2010**, 89, 1796–1805.

26. American Society for Testing and Materials (ASTM). *ASTM D2710. Standard Test Method for Determination of Bromine Index of Petroleum Hydrocarbons by Electrometric Titration*. ASTM: West Conshohocken, PA, 2011.

27. American Society for Testing and Materials (ASTM). *ASTM D1159. Standard Test Method for Determination of Bromine Numbers of Petroleum Distillates and Commercial Olefins by Electrometric Titration*. ASTM: West Conshohocken, PA, 2011.

28. Ruzicka, D. J.; Vadum, K. Modified method measures bromine number of heavy fuel oils. *Oil Gas J.*, **1987**, 85(31), 48–50.

29. Unger, E. H. Influence of olefin structure on bromine number as determined by various analytical methods. *Anal. Chem.*, **1958**, 30(3), 375–380.

30. Lubeck, A.; Cook, R. D. Bromine number should replace FIA in gasoline olefins testing. *Oil Gas J.*, **1992**, 90(52), 86–89.

31. Ceballo, C. D.; D'Ambrosio, F.; Torres, N. Estimation of olefins to aromatics ratio (O/A) in cracked naphthas by bromine number assay. *Pet. Sci. Technol.*, **1998**, 16(1–2), 179–189.

32. Briker, Y.; Ring, Z.; Yang, H. Integrated methodology for characterization of petroleum samples and its application for refinery product quality modeling, Chapter 4 In: *Practical Advances in Petroleum Processing*, Hsu, C. S., Robinson, P. R. (Eds.). Springer: New York, 2006; pp. 117–147.

33. Galarraga, C.; Pereira-Almao, P. Hydrocracking of Athabasca bitumen using submicronic multimetallic catalysts at near in-reservoir conditions. *Energy Fuels*, **2010**, 24, 2383–2389.

34. Loria, H.; Trujillo-Ferrer, G.; Sosa-Stull, C.; Pereira-Almao, P. Kinetic modeling of bitumen hydroprocessing at in-reservoir conditions employing ultradispersed catalysts. *Energy Fuels*, **2011**, 25, 1364–1372.

35. Fathi, M. M.; Pereira-Almao, P. Catalytic aquaprocessing of Arab Light vacuum residue via short space times. *Energy Fuels*, **2011**, 25, 4867–4877.

36. Carbognani, L.; Meneghini, R.; Hernandez, E.; Lubkowitz, J.; Pereira-Almao, P. Applications of group-type and class-type analysis via simulated distillation-mass spectrometry for process upgrading monitoring. *Energy Fuels*, **2012**, 26(4), 2248–2255.

37. Orrego-Ruiz, J. A.; Mejia-Ospino, E.; Carbognani, L.; Lopez-Linares, F.; Pereira-Almao, P. Quality prediction from hydroprocessing through infrared spectroscopy (IR). *Energy Fuels*, **2012**, 26(1), 586–593.

38. Carbognani, L.; Gonzalez, M. F.; Pereira-Almao, P. Characterization of Athabasca vacuum residue and its visbroken products. Stability and fast hydrocarbon group-type distributions. *Energy Fuels*, **2007**, 21(3), 1631–1639.

39. Carbognani, L.; Lubkowitz, J.; Gonzalez, M. F.; Pereira-Almao, P. High temperature simulated distillation of Athabasca vacuum residue fractions. Bimodal distributions and evidence for secondary "on-column" cracking of heavy hydrocarbons. *Energy Fuels*, **2007**, 21(5), 2831–2839.

40. Pereira-Almao, P.; Trujillo, G.; Peluso, E.; Galarraga, C.; Sosa, C.; Scott Algarra, C.; Lopez-Linares, F.; Carbognani Ortega, L.; Zerpa Reques, N. Systems and methods for catalytic steam cracking of non-asphaltene containing heavy hydrocarbons. Canadian Patent Application CA 2781192 (Filed 2012-06-28. Open to Public Inspection on December 30, 2012.).

41. Boyland, E. The biological significance of metabolism of polycyclic compounds. *Biochem. Soc. Symp.*, **1950**, 5, 40–54.

42. Grover, P. L.; Hewer, A.; Sims, P. Metabolism of polycyclic hydrocarbons by rat-lung preparation. *Biochem. Pharmacol.*, **1974**, 23, 323–332.

43. Sims, P.; Grover, P. L.; Swaisland, A.; Pal. K.; Hewer, A. Metabolic activation of benzo[a]pyrene proceeds by a diol-epoxide. *Nature*, **1974**, 252, 326–328.

44. Strausz, O. P.; Lown, E. M. *The Chemistry of Alberta Oil Sands, Bitumen and Heavy Oils*. AERI Editions: Calgary, Alberta, 2003.
45. Kushnarev, D. F.; Polonov, V. F.; Donskikh, V. I.; Kiryukhina, E. D.; Kalabin, G. A. Methods of analysis ^{1}H NMR spectroscopy determination of olefin hydrocarbons in petroleum products. *Chem. Technol. Fuels Oils*, **1991**, 27(6): 334–336.
46. Process NMR Associates. Available at http://process-nmr.com/WordPress??p=42 (downloaded June 2012).
47. Story, R. L.; Provine, R. W.; Bennett, H. T. Chemistry of gum formation by cracked gasoline. *Ind. Eng. Chem.*, **1929**, 21(11), 1079–1084.
48. Jones, L.; Hazlett, R. N.; Li, N. C. Storage stability studies of fuels derived from shale and petroleum. *Prep. Papers Am. Chem. Soc. Div. Fuel Chem.*, **1983**, Seattle 03–83, pp. 196–201.
49. Beal, E. J.; Cooney, J. V.; Hazlett, R. N.; Morris, R. E.; Mushrush, G. W.; Beaver, B. D.; Hardy, D. R. *Mechanisms of Syncrude/Synfuels Degradation*. U.S. Department of Energy, Washington, DC. Report DOE/BC/10525-16, April 1987; 134 pp.
50. Watkinson, A. P.; Li, Y. H. Fouling characteristics of a heavy vacuum gas oil in the presence of dissolved oxygen. In: *Proceedings of the International Conference on Heat Exchangers Fouling and Cleaning VIII* (peer-reviewed), Muller-Steinhagen, H., Malayeri, M. R., Watkinson, A. P. (Eds.). PP Publico Publications, Schladming, Austria, 2009; pp. 27–32.
51. American Society for Testing and Materials (ASTM). *ASTM D525-05. Standard Test Method for Oxidation Stability of Gasoline (Induction Period Method)*. ASTM: West Conshohocken, PA, 2011.
52. American Society for Testing and Materials (ASTM). *ASTM D873-02(2007). Standard Test Method for Oxidation Stability of Aviation Fuels (Potential Residue Method)*. ASTM: West Conshohocken, PA, 2011.
53. American Society for Testing and Materials (ASTM). *ASTM D381. Standard Test Method for Gum Content in Fuels by Jet Evaporation*. ASTM: West Conshohocken, PA, 2011.
54. Carbognani, L.; Lopez-Linares, F.; Trujillo, M.; Wu, Q.; Carbognani, J.; Pereira-Almao, P. On the determination of olefins in heavy oil, bitumen and their upgraded products. *Prepr. Papers Am. Chem. Soc. Div. Petrol. Chem.*, **2012**, 57(2), 685–688.
55. Carbognani, L.; Lopez-Linares, F.; Trujillo, M.; Wu, Q.; Carbognani, J.; Pereira-Almao, P. Olefin content estimation in heavy oil, bitumen and their upgraded products. *Prepr. Papers Am. Chem. Soc. Div. Energy Fuels*, **2013**, 57(2), 507–508.
56. Silva, A. C. O.; SanGil, R. A. S.; Kaiser, C. R.; Azevedo, D. A.; Dávila, L. A. Combining NMR and GC-MS to characterize olefin rich fractions of automotive gasolines. *Ann. Magn. Res.*, **2006**, 5(1/3), 11–21.

Section IV

Analytical Measurements

6 Advances in Gas Chromatography for Petroleum Upstream, Refining, Petrochemical, and Related Environmental Applications

Carl E. Rechsteiner Jr., John Crandall, and Ned Roques

CONTENTS

CONTEXT

Gas chromatography (GC) is a primary measurement tool for understanding petroleum composition throughout the value chain, from exploration through refining and fuel marketing. Changes in GC technology allow

- Faster and robust measurements from the laboratory to the process line
- Greener operation; lower power and cooling requirements, reduced solvent volumes, and higher throughput per unit time

ABSTRACT

Gas chromatography (GC) continues to be a workhorse tool for the oil and petrochemical industries. Until recently, research-grade gas chromatographs were needed to meet the data quality objectives for exploration, production, refining, and marketing operations. Advances in all aspects of the GC process, inlets, columns, detectors, heating, etc., allow near-research-grade performance in small modular and highly configurable systems. This chapter discusses the requirements, capabilities, and application of such systems useful for critical measurements in the oil and petrochemical industries. Included are examples for gas analyses, target compound (process indicator and environmental) analyses, and simulated distillation in laboratory, process, and transportable environments.

6.1 HISTORICAL OVERVIEW OF USE

Gas chromatography (GC) is ideally suited for mixtures, even complex mixtures as long as they are gases or liquids that can be vaporized at <350°C. It is extremely flexible and can be configured to perform a variety of measurements. Semiqualitative and analytically quantitative information results in both good information about what the material for analysis contains, and in many cases precisely how much of the target analyte is present. Very early in the development of GC as a research tool, it quickly became the "darling" in the analysis of petroleum products. Boiling range distributions and specific component analyses that in many ways emulate the techniques used to make petroleum products, such as distillation, were familiar to the engineers, scientists, and technicians requiring compositional data to keep the process on specification.

However, the GC technology and the users of the data have long been involved in a love–hate relationship. Most just need and want the answers that GC can provide for process operations. They just want the data and the confidence that the data is valid. They do not necessarily want a gas chromatograph. The batch nature of the technique is often perceived as slow, requires a number of utilities (gases, electricity, electronics, and computing), has very high safe-installation costs, may require too much maintenance, and has been viewed as "too fussy."

Many have attempted to displace the GC with spectroscopic or other analytical techniques. In the end, most discover that just as the carpenter builds a house using a variety of tools, the analyst must match the tool to the job at hand. For petroleum analyses, the gas chromatograph is as important to the analyst as the hammer is to the carpenter. One should not try to drive a nail with a saw nor should one try to measure boiling range distributions with a spectrometer.

6.2 PRACTICE OF GC

A gas chromatograph is an electronic, analytical instrument useful for gas and liquid sample mixtures employing separation technology for both qualitative analysis (what is it?) and quantitative analysis (how much of it is present?). A gas chromatograph is singularly capable of isolating components of interest from interferences demonstrating boiling range distribution and establishing compositional properties of sample mixtures from air components up to virtually any hydrocarbon that can be vaporized at <350°C.

GC is intuitively understandable and provides expert results without the need for expert users in the vast majority of the applications. GC is a batch process

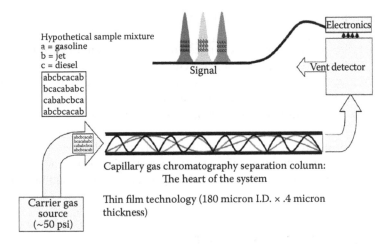

FIGURE 6.1 Simple picture of how a gas chromatograph works.

consisting of three fundamental steps: sampling, separation, and detection (refer to Figure 6.1).

In sampling, a precise and representative quantity of the sample mixture is isolated with a syringe or automated valve assembly. Chromatographic sample sizes range from 60 nL to 100 μL, and in some specialized cases up to milliliter sizes. Gases and liquids can be measured as long as the liquid can be vaporized at temperatures below the point where decomposition occurs (for hydrocarbons, usually <350°C).

Once the sample is obtained, it is introduced into a flowing stream (carrier or mobile phase) delivering the sample to the separation column. Physical or chemical interactions between the column media (stationary phase, typically a bonded thin film of nonpolar to polar hydrocarbon liquid), the sample, and the carrier gas results in a sorting of molecular types in the sample. Temperature control can be either isothermal or programmed to increase with analysis time continuously or in stages. Proper choice of column and operating conditions (flow rate and temperature) results in separations to enable interference-free measurement of the desired components of the mixture. As the analysis cycle progresses, separated components elute from the column and progress into the detector.

As the separated components begin to elute from the column, the analyst, using the appropriate electronics, is able to "see" individual groups or peaks of specific components. Detectors are designed to use a property of the components of interest that is different from the carrier properties to generate an electronic signal proportional to the amount of the components. Detectors can be universal responding to all components (thermal conductivity detector [TCD]) or specific responding to hydrocarbons (flame ionization detector [FID]), or even more specific (flame photometric detector) responding to sulfur-containing components. The retention time of the component is qualitative (what is it?), while the area under the curve is quantitative (how much is it?). Calibration samples are used to relate known component area to the concentration of the same component in unknown samples.

In most applications, the utility requirements are

- Clean, dry, hydrocarbon-free compressed air for pneumatic switching, purge gas, and oxidant for flame-based detectors
- Carrier gases, usually helium, nitrogen, or hydrogen that is preferable now more than ever because of the cost of helium
- Electricity often as much as 3 kW per gas chromatograph

While these utilities are not too difficult in a research laboratory, they can become quite difficult and expensive in routine use whether in a plant support laboratory, transportable applications, and certainly in process installations for online analysis.

6.3 MICRO-GC AND ITS IMPACT

Many attempts have been made since the early 1980s to mitigate the "hate" part of the love–hate relationship between gas chromatographs and the end consumer of the data and information a gas chromatograph can provide. The micro gas chromatograph was conceived on the basis of micromachining. Some called these early devices "chromatograph on a chip." Really the plumbing for the chromatograph was on the chip; the chromatographic components (sample valve, column, and detector) were mounted onto the chip.

These versions were conceived to reduce size, utility consumption, analysis time, and many of the "hate" drivers for the GC. Taking these steps for GC while retaining chromatographic system performance is not a simple scaling issue. As each component of the system is scaled down, attention must be paid to minimize the impact of physical differences such as dead volumes and small flows, control issues such as consistency in column heating and temperature profiles, and the software control of the system electronics to permit precise changes to the GC conditions.

One of the most important issues in retaining the performance is the ability to achieve temperatures suitable for hydrocarbon measurements up to the limits for hydrocarbon stability. The research-grade instruments achieve these temperatures using high thermal mass components, bulky air bath ovens for the columns, and consumption of massive amounts of energy. Historically, most so-called micro gas chromatographs have been severely limited by the temperatures attainable with micromachined, chip-based techniques. There has simply been no way to heat components to the required temperatures for more than the simplest applications involving air components and low boiling hydrocarbons up to about C_6.

Thus, something more has been sought, requiring a rethinking about the definition of GC in general and especially about micro-GC. This definition for micro-GC was proposed as part of a new UltraFast ASTM method effort.

A micro gas chromatograph is a GC resulting from the elimination of the large, high thermal mass column oven and downsizing the GC components from millimeter to micrometer dimensions incorporating resistively heated steel chromatography columns resulting in analysis times 10–50 times faster than conventional GCs.

This definition was rejected as being "competitively restrictive." However, deploying this definition by throwing out both the high thermal mass design constraints of the past and learning from the micromachined technique experience, it became possible to bring a "micro gas chromatograph without the limitations of a micro gas chromatograph" to the forefront of GC technology.

The method published in January of 2013 as ASTM D7798 is entitled "Standard Test Method for Boiling Range Distribution of Petroleum Distillates with Final Boiling Points up to 538°C by Ultra Fast Gas Chromatography (UFGC)." The clear benefits from the older but still in use ASTM D2887 method are speed, precision, instrument size, and many more. However, the largest single benefit of rethinking micro- and fast GC is temperature. The final boiling point of 538°C, equivalent to n-C_{44}, brings micro-GC capabilities to the petroleum industry for the first time.

It has been estimated that about 80% of the required measurement need in the petroleum industry can be covered for samples having end boiling points of <538°C. In exploration and production, GC complements other measurements by focusing on the oil composition for down-hole samples (both gases and liquids), drilling core extracts, drill stem tests, crude assays, reservoir continuity, and biomarker analysis. The measurement results for gas condensates and oils include gas composition, British thermal unit (BTU) content, and crude oil fingerprinting and assay, all directed at value assessments and their use in custody transfer of the petroleum products.

In refining and marketing, the gas chromatograph is routinely used for process monitoring and control, diagnostics during process upset, product release, and ensured protection of the environment. While the list of possible applications is very long, a general description of applicability includes refinery gas analysis, detailed composition of the light streams, yield curve determination for feedstock utilization, fingerprinting of process operations for optimization, and sulfur speciation of the products. GC is often used to resolve product integrity disputes and environmental assessment.

Figure 6.2 shows the vast scope of GC applicability in the refining process alone. A fast, micro gas chromatograph that can be configured for air components, gas-phase or low-boiling hydrocarbons, middle distillates, and even fingerprinting of complex liquids up to n-C_{44} should be extremely useful for not only refining but also exploration, production, transmission, distribution, and marketing of petroleum products and petrochemicals.

A gas chromatograph that is easier to use, smaller and lighter, smarter with built-in advanced data processing, faster, and having a much lower utility requirement would mitigate much of the "hate" aspects for the data consumers, and not only open the door for continued use in legacy applications but should also lead to dramatic expansion in areas where previous techniques simply could not be economically or safely deployed. Most end data consumers say that to expand the use of GC in petroleum and petrochemical process control applications, the analysis cycle time must be reduced to a few minutes, and the footprint and infrastructure requirements of space, power, gases, etc., must also be reduced, all while maintaining the performance of research-grade GCs.

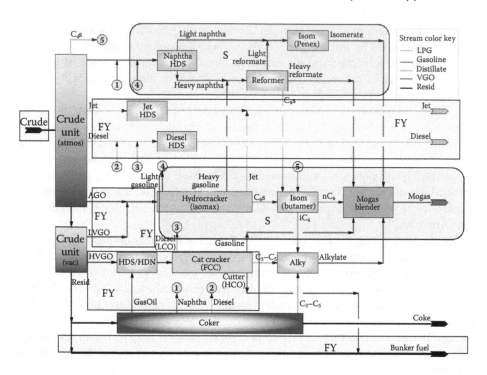

FIGURE 6.2 Scope of GC applications in the refinery. GC analyses that can identify specific species are in the shaded area labeled S. GC analyses that provide fingerprints and yield curves but cannot identify specific species are in the shaded area labeled FY.

6.4 INSTRUMENTAL AND EXPERIMENTAL EXAMPLE FOR MICRO- AND FAST GC DEPLOYMENT

The Calidus micro gas chromatograph and fast gas chromatograph from Falcon Analytical were used for all GC analyses. The Chromperfect 7 software was used for GC data acquisition and processing. The chromatographic data alignment software LineUp, from Infometrix, was used to improve chromatographic repeatability, if needed.

Bringing forward a "micro gas chromatograph without the limitations of a micro gas chromatograph" while retaining chromatographic system performance makes giant strides to improving acceptance. However, this alone does nothing for the perception of complexity, maintenance problems, and installation costs. Thus, the same attention to detail must be applied to instrumental implementation.

This attention is evidenced in the modular design of the gas chromatograph. Basic modules for the injector, detector, and columns are swappable to give a variety of configurations depending on the analysis needs. For the academic setting, this allows maximum flexibility in a single unit. For the industrial setting, it minimizes instrumental downtime leading to increased throughput both from the inherent speed of analyses and the reduced instrument repair time. Furthermore, the system requirements, in terms of size, weight, and modularity, make the gas chromatograph usable in the laboratory, in the field (enclosures suitable for class 1, division II in some regions, zone 1 or zone 2 in others), and as a transportable unit to be taken to the point of interest.

The petroleum samples analyzed were obtained from a variety of sources. These samples include crude oils, distilled fractions from crudes, and refinery and plant products from gaseous to highly viscous. In other words, the samples analyzed ranged from start (5°C above ambient) to 538°C (for downstream applications) or permanent gases and light hydrocarbons (in natural gas) and C_1 to C_{44} (for upstream applications).

6.5   RESULTS AND DISCUSSION

Gas analyses are critically important throughout the petroleum industry value chain. The price of natural gas depends on the composition of the gas. From the explicit component identification and amount, one can compute the BTU content of the gas and identify any "bad actors," such as H_2S and CH_4S, which would lower the price from the producer. Similarly, gas measurements for refinery salable products, e.g., propane and butane, have similar pricing constraints on BTU content and "bad actor" concentrations. Furthermore, gas analysis is widely used for process control in refining and chemical plants.

To illustrate the fast GC and micro-GC performance, an example of an extended natural gas analysis is shown. This analysis uses a column-switching valve to deliver the sample to two independent column modules. A nonpolar column is combined with an FID to measure hydrocarbons out to about C_{10}, while a molecular sieve column module is combined with a TCD to measure permanent gases such as O_2 and N_2.

Figure 6.3a and b show typical chromatograms for hydrocarbons and permanent gases from a single sample injection. Hydrocarbons shown in Figure 6.3a and b include both normal- and iso-paraffins through C_{10} with good separation. The ratio of normal- to iso-paraffins provides useful refinery diagnostics, e.g., for optimum reformer operation. Figure 6.4 shows a chromatogram with the TCD detector where CO_2, O_2, N_2, and water are clearly seen. As expected for a molecular sieve column, the water peak shows significant tailing.

The modular system also permits heart cutting, backflushing, and as shown in the use in the natural gas calibration chromatograms, the trap/bypass configuration to enhance both chromatographic separation and trace analyses. Figure 6.5 shows the four basic configurations of the modular gas chromatograph. Variations on each of these enable virtually eliminating any limitations for the methods development chromatographer. Figure 6.6 shows the sample processing module (main chassis), column (1–8 m per module for up to 16 m total), and detector modules. Example configurations are shown.

GC of crude oils is a key technology in exploration and for the valuation of crude oil. Figure 6.7 shows the chromatogram for a crude oil sample obtained from a drill stem test. Key features include the prominent n-paraffin peaks that permit assessing the crude oil on the basis of its carbon number, as well as the measurement of pristane and phytane whose ratio is useful in assessing crude oil maturation.

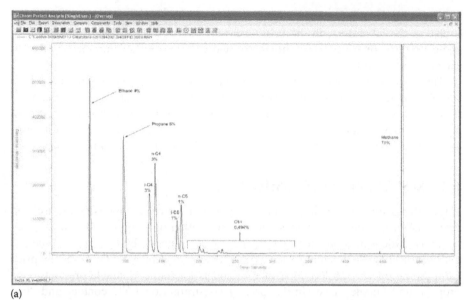

(a)

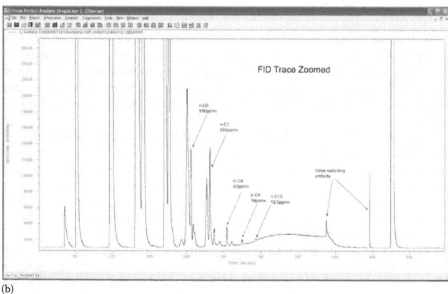

(b)

FIGURE 6.3 (a) FID trace showing extended natural gas calibration standard to *n*-C10. (b) *Y*-axis expansion of FID trace extended natural gas calibration standard to *n*-C10.

GC technology is widely used in refining to provide yield distribution information on feeds and products to refinery processes. Simulated distillation as embodied by ASTM Method D-2887 and its variants provides this information. Table 6.1 and Figure 6.7 provide information on the use of this micro gas chromatograph as part of the development of the Ultrafast D-2887 method (see page 117).

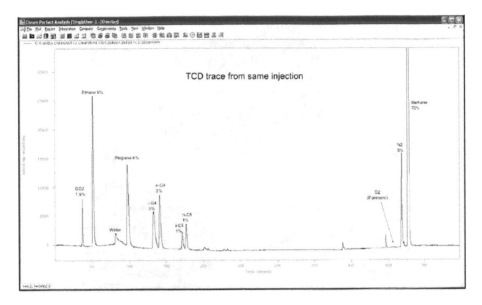

FIGURE 6.4   TCD trace extended natural gas calibration standard to *n*-C10.

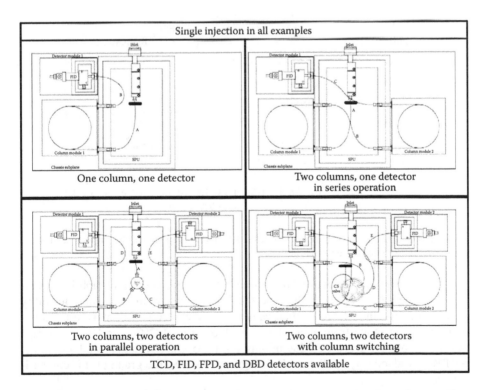

FIGURE 6.5   Basic modular GC configurations. TCD, thermal conductivity detector; FID, flame ionization detector; FPD, flame photometric detector; DBD, dielectric barrier detector.

FIGURE 6.6 Trans-configurable modular chromatographic assembly. US Patent 8336366.

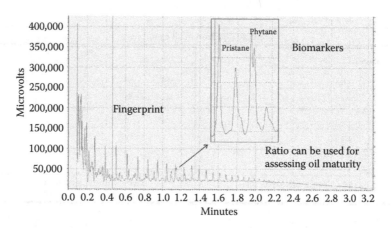

FIGURE 6.7 Crude oil from drill stem tests. Expanded area shows the important pristane and phytane region of the chromatogram.

Table 6.1 shows the data obtained from a micro gas chromatograph in a refinery plant support laboratory. The repeatability is excellent and the reproducibility is comparable to the ASTM method values for a reference gas oil standard that has a boiling point range from about 200°F to 900°F (93–482°C) and has an average deviation from the accepted value of only 1°F (0.55°C). The data is well within the ASTM reproducibility for the method. Figure 6.8 shows the graph of this comparison data.

These examples demonstrate the use of a micro gas chromatograph with near-research-grade performance for the analysis of petroleum and petrochemicals.

TABLE 6.1

Fifteen Replicate Campaigns of a Refinery Plant Support Laboratory

Rep #	0.5%	5%	10%	15%	20%	25%	30%	35%	40%	45%	50%	55%	60%	65%	70%	75%	80%	85%	90%	95%	99.5%
1	241.3	304.6	349.1	394.8	436.5	471.3	500.0	527.3	553.5	577.5	594.6	610.7	629.3	648.7	668.6	690.1	712.8	737.2	765.3	804.4	885.6
2	240.5	304.4	349.1	394.9	436.8	471.3	500.3	527.7	553.6	577.7	595.0	611.1	629.7	649.3	669.1	690.6	713.3	737.7	766.1	805.3	886.9
3	241.0	304.4	349.2	394.7	436.8	471.3	500.5	527.8	553.5	577.5	594.6	610.7	629.1	648.8	668.5	690.3	712.8	737.0	765.3	804.6	885.7
4	240.5	304.5	349.1	394.9	437.0	471.4	500.4	527.7	553.7	577.6	594.7	610.9	629.3	648.9	668.6	690.5	712.9	737.2	765.7	804.9	888.8
5	240.9	304.4	349.3	395.0	437.1	471.6	500.4	527.7	553.9	577.6	594.8	610.7	629.3	648.7	668.6	690.2	712.6	737.0	765.5	804.9	886.2
6	240.6	304.3	349.0	394.6	436.7	471.2	500.2	527.3	553.4	577.3	594.4	610.5	629.0	648.7	668.4	690.0	712.6	736.8	765.2	804.7	887.6
7	240.7	304.4	349.2	394.8	436.7	471.2	500.0	527.3	553.3	577.4	594.5	610.4	629.0	648.5	668.3	689.8	712.4	736.7	765.0	804.0	886.8
8	239.5	304.1	349.1	395.1	437.3	471.6	500.4	527.5	553.6	577.3	594.6	610.4	628.9	648.5	668.3	689.9	712.3	736.6	765.1	804.4	885.5
9	240.5	304.5	349.3	394.9	436.9	471.5	500.5	527.6	553.6	577.3	594.6	610.5	629.1	648.7	668.7	690.4	713.0	737.2	765.3	804.4	885.8
10	240.8	304.6	349.4	395.1	437.3	471.8	500.8	528.0	553.8	577.6	595.0	611.1	629.5	649.2	668.9	690.5	713.1	737.0	765.1	804.7	887.7
11	240.8	304.4	349.4	394.8	437.1	471.7	500.7	527.8	554.0	577.7	595.0	611.1	629.7	649.3	668.9	690.4	712.8	736.6	765.1	804.4	885.4
12	240.9	304.5	349.1	394.9	437.0	471.5	500.4	527.6	553.4	577.4	594.6	610.4	629.1	648.5	668.3	689.8	712.4	736.8	764.9	803.8	885.0
13	241.0	304.6	349.4	395.3	437.3	472.0	500.9	528.1	554.0	577.6	594.8	610.5	629.0	648.5	668.3	689.8	712.4	736.8	764.9	804.0	885.4
14	241.0	304.5	349.1	394.9	436.8	471.4	500.5	527.8	553.8	577.7	595.0	610.7	629.6	649.0	668.8	690.5	713.0	737.4	766.0	805.2	886.7
15	240.7	304.5	349.4	395.2	437.6	472.1	501.1	528.1	553.8	577.5	594.7	610.7	629.0	648.9	668.6	690.4	712.9	737.4	765.7	805.4	888.4
Ave.	240.7	304.5	349.2	394.9	437.0	471.5	500.5	527.7	553.6	577.5	594.7	610.7	629.2	648.8	668.6	690.2	712.7	737.1	765.3	804.6	886.5
SDev	0.39	0.12	0.13	0.19	0.28	0.27	0.29	0.24	0.22	0.14	0.20	0.25	0.25	0.27	0.24	0.27	0.30	0.31	0.39	0.47	1.13
RSD	0.16%	0.04%	0.04%	0.05%	0.07%	0.06%	0.06%	0.05%	0.04%	0.02%	0.03%	0.04%	0.04%	0.04%	0.04%	0.04%	0.04%	0.04%	0.05%	0.06%	0.13%
Consensus	239	304	349	393	435	469	499	526	552	576	594	610	629	649	669	690	712	736	764	803	887
Difference	1.71	0.45	0.21	1.94	1.99	2.53	1.47	1.69	1.64	1.52	0.73	0.72	0.24	−0.19	−0.41	0.22	0.75	1.06	1.35	1.59	−0.50

Note: Initial BP = 241°F; Final BP = 886°F; Ave. SDev = 0.3°F; Ave. RSD = 0.05%; Ave. difference from consensus value = 1.0°F.

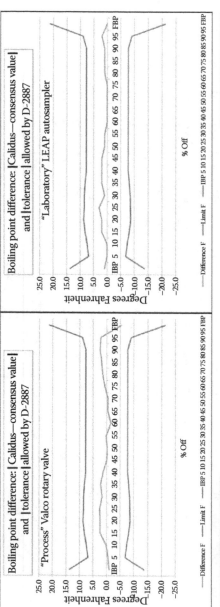

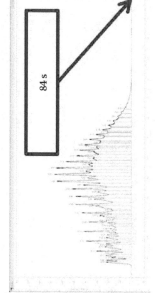

FIGURE 6.8   Comparison of results to reference gas–oil consensus value from process and laboratory systems.

7 Application of NMR Technology in Petroleum Exploration and Characterization

Zheng Yang, Ajit Pradhan, John Zintsmaster, and Boqin Sun

CONTENTS

CONTEXT

- NMR technology is a game-changing technology for petroleum exploration and production
- Many applications have been found in formation evaluation and reserve characterization

ABSTRACT

The application of nuclear magnetic resonance (NMR) technology in petroleum exploration and characterization dates back to the early 1950s, leading to the development of the NMR logging tool in the early 1960s. However, the logging tool had many limitations and was eventually retired from use. It took the petroleum industry nearly 40 years of continual efforts to develop cutting-edge NMR technology for upstream applications. Today, the petroleum industry uses NMR technology for hydrocarbon reserve estimation, production efficiency determination, and reservoir management. Such a wide variety of valuable information obtained using NMR technology has had a tremendous impact on and provided economic benefits to the petroleum industry. In this chapter, we will summarize how NMR technology is used today to study the properties of hydrocarbons and formation rocks, such as mineralogy-independent porosity, irreducible water saturation, permeability, and viscosities. The major applications of NMR technology in the petroleum industry will be briefly introduced and discussed with examples.

7.1 INTRODUCTION

Nuclear magnetic resonance (NMR) refers to the response of atomic nuclei to magnetic fields. In the presence of an external magnetic field, an atomic nucleus precesses around the direction of the external field in much the same way a gyroscope precesses around the earth's gravitational field. Measurable signals can be produced when these spinning magnetic nuclei interact with the external magnetic fields [1]. Since its discovery in the late 1940s, NMR spectroscopy has become the preeminent technique to determine the structure of organic compounds. In the early 1950s, NMR technology was introduced to the studies on the properties of brine and/or oil-saturated rock samples. It soon became clear that the local heterogeneous field of the fluid-saturated rocks causes such a large line broadening that it was impossible to resolve the oil and water signals. One would have to resort to the difference in the relaxation rates between oil and water to resolve the signals. Thus, the relaxation times of oil and water are exploited by NMR technology.

The first generation of NMR logging tools for oil exploration was built in the 1950s using the earth's magnetic field. However, these tools did not gain wide popularity because of their numerous limitations. In the early 1990s, the invention of NMR logging tools with permanent magnets and pulsed radiofrequency signals provided the petroleum industry with powerful new methods for evaluating petroleum reservoirs [2,3]. Rather than placing the sample at the center of the instrument, the NMR logging tool turns the laboratory NMR equipment inside–out and places itself in a well bore, and is surrounded by the formation to be analyzed. A permanent magnet is placed inside the NMR logging tool to produce a magnetic field that polarizes the formation materials. An antenna is incorporated to surround this magnet, which is used to perturb the spins, and then "listen" for the decaying echo signal from those protons that are in resonance with the field from the permanent magnet.

Theoretically, NMR measurements can be made on any nucleus that has an odd number of protons or neutrons, or both, such as the nucleus of hydrogen (^{1}H), carbon (^{13}C), and sodium (^{23}Na). For most of the elements found in earth formations, the nuclear magnetic signal induced by external magnetic fields is too small to be detected with a borehole NMR logging tool. However, hydrogen, which has only one proton and no neutrons, is abundant in both water and hydrocarbons, has a relatively large gyromagnetic ratio, and produces strong signals. To date, almost all NMR logging and NMR rock studies are based on the responses of the nucleus of the hydrogen atom.

In recent years, as the conventional oil reserves in the world continue to decline and the worldwide demand for oil continue to increase, unconventional oil deposits attracted increasing attention from the industry and are considered to be an important part of the future of world energy security. Numerous companies have invested billions of dollars in oil sand surface mining and in situ recovery projects for heavy oil. Furthermore, as the oil price keeps increasing, the exploration of bitumen in deeper formations is drawing more and more interest. Many in situ heavy oil recovery options have been designed and commercialized for recovering oil in the deeper formations [4]. Therefore, being able to predict heavy oil properties and fluid saturation in situ, as well as to optimize the heavy oil recovery process, is of considerable value to the industry. NMR technology has also been shown to have great potential for the exploration and characterization of heavy oil reservoirs [5–7].

7.2 NMR RELAXATION IN POROUS MEDIA

In NMR, the relaxation describes several processes by which nuclear magnetization prepared in a nonequilibrium state returns to the equilibrium. The longitudinal relaxation time (T_1), transverse relaxation time (T_2), and self-diffusion coefficient (D) of fluids are the three major NMR parameters of the interest to the petroleum industry. For fluid-saturated porous media (e.g., formation rocks), the relaxations consist of three independent mechanisms:

- Relaxation of bulk fluid, which affects both T_1 and T_2 values
- Relaxation caused by surface relaxations, which affects both T_1 and T_2 values
- Relaxation enhanced by the effect of spin diffusion in a field gradient, which only affects the T_2 relaxation rate

Therefore, the apparent T_2 and T_1 relaxation times of the fluid-saturated porous media can be written in the following forms:

$$\frac{1}{T_2} = \frac{1}{T_{2B}} + \frac{1}{T_{2S}} + \frac{1}{T_{2D}}$$
(7.1)

$$\frac{1}{T_1} = \frac{1}{T_{1B}} + \frac{1}{T_{1S}}$$
(7.2)

Here, T_{1B} and T_{2B} represent the T_1 or T_2 value for the bulk fluid; T_{1S} and T_{2S} represent the relaxation caused by the surface relaxivity; and T_{2D} represents the T_2 relaxation caused by the diffusion effect.

The relaxation behavior of fluid-saturated porous media was first studied by Brownstein and Tarr [8]. They demonstrated that the magnetization of such a system can be described by multiexponential decay, which arises simply as a consequence of diffusional spins in porous media of various sizes and shapes. They showed that if the porous media has a distribution of pore sizes, then there would be a distribution of relaxation times. Since then, numerous correlations have been developed to correlate the NMR signals to the physical or chemical properties of either formation rocks or reservoir fluids.

7.3 NMR LOG DATA ACQUISITION AND INVERSION

Usually after seismic prospecting, which determines specific areas with particular geological structures that are likely to have oil deposits, exploratory wells are often drilled. There are various kinds of logging tools developed in the petroleum industry that can be lowered to exploratory wells to measure certain properties of the fluid-saturated earth formation to estimate the petroleum reserve. Historically, resistivity logging has been the primary method, albeit an indirect approach, for estimating hydrocarbon reserves. However, the resistivity logging tools are strongly influenced by the presence of conductive materials in the formation. In comparison, NMR technology has a distinctive advantage, as the NMR logging tool directly measures the density of hydrogen nuclei in the reservoir fluids and gives the most reliable estimates of the hydrocarbon reserve. The California Research Company (now Chevron Energy Technology Company) and Byron Jackson Tools introduced the first NMR logging tool in the 1950s [9]. This logging tool was really not much more than a prototype and suffered from reliability and interpretation problems. After nearly 40 years of continued developmental efforts, NMR technology has matured, and today the new NMR-based tools can operate at multiple frequencies, allowing independent information from multiple zones, faster speeds, improved signal-to-noise ratio, and different pulse-timing sequences for complex data acquisitions.

Normally, NMR measurements are carried out right after the exploratory well is drilled and while the tool is being pulled upward from the bottom of the well. Since measuring the relaxation time T_2 is much faster, the data acquisition schemes of Carr–Purcell–Meiboom–Gill (CPMG) echo trains are usually acquired during the logging operation [10,11]. There are various methods of inverting the CPMG echo train data to obtain a T_2 distribution. Sun and Dunn [12] introduced two NMR inversion methods within the framework of one-dimensional (1D) NMR by varying the echo spacing and wait time. These methods are similar in principle to the time domain approaches; however, the mathematical manipulation is different where a shift matrix scheme in the T_2 domain is used to incorporate the diffusion effect. The first method, FET (fluid typing by editing T_2 distribution), inverts the T_2 distribution of each fluid from the apparent T_2 distribution using a shift matrix. It offers a workable method in the T_2 domain. The second method, GIFT (global inversion of fluid typing), couples the T_2 distribution of each fluid directly with the CPMG echo train through a global matrix E_G.

The 1D NMR measurement, which usually measures T_2 relaxation times, is still the most common application of NMR technology in the petroleum

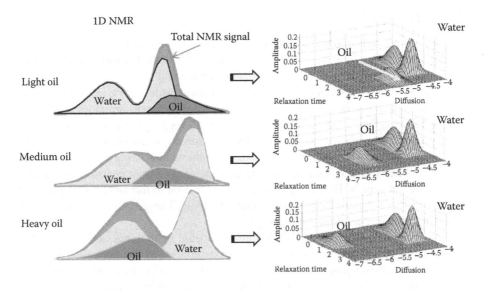

FIGURE 7.1   Comparison between 1D NMR and 2D NMR for different oils.

industry. However, in some scenarios (particularly in heavy oil reservoirs), the accuracy of fluid typing is greatly overshadowed by the overlap between the T_2 peaks arising from different fluids with similar apparent T_2 relaxation times. To overcome this limitation, the 2D NMR technique was introduced to NMR logging in oil exploration and reservoir characterization [13–16]. In 2D NMR measurement, the proton population distribution is not only a function of T_2 relaxation times but of the fluid diffusion coefficient D or relaxation time T_1. As illustrated in Figure 7.1, it is easier to distinguish the oil signals from water signals in a 2D NMR map and to perform accurate quantitative analysis of different reservoir fluids, owing to the greater differences in diffusion coefficients or T_1 relaxation times for different fluid components [17].

   In the following, we will illustrate some of the major applications of NMR technology in the petroleum industry for exploration and formation characterization.

7.4   APPLICATION OF NMR TECHNOLOGY IN PETROLEUM EXPLORATION

Since the introduction of the pulsed NMR logging tools, investigators have found numerous applications of NMR technology in formation evaluation. The most important ones are mineralogy-independent porosity estimation, irreducible water determination, permeability prediction, and viscosity determination. No other logs provide similar kinds of information [3]. These applications are summarized in the following.

7.4.1 Porosity Estimation

Porosity estimation is important because a large porosity of the earth formation indicates a potential of having large hydrocarbon reserves. Traditionally, porosity is measured by conventional logging tools such as neutron, bulk-density, and acoustic-travel-time porosity logging. However, conventional logging tool measurements can be influenced by the components of reservoir rocks [18]. Therefore, conventional tools are more sensitive to matrix materials than to pore fluids. In contrast, NMR logging tools primarily measure the total population of hydrogen nuclei in the pore space containing water and oil, and is quite insensitive to the hydrogen nuclei in the solid matrix owing to their extremely long T_1 and short T_2 [19]. Consequently, matrix materials do not contribute to porosities measured by an NMR logging tool so that calibration to formation lithology is not necessary. This response characteristic makes NMR well logging tools fundamentally different from other conventional logging tools.

The initial amplitude of the NMR spin-echo train or the area under the T_2 distribution curve is proportional to the number of protons of the pore fluids present within the sensitive volume. The amplitude of the spin-echo train can be calibrated to give a porosity value for any earth formation, as shown in Equation 7.3:

$$\phi = \sum_{j=1}^{N} a_j \tag{7.3}$$

where ϕ is the porosity of formation.

However, to decrease the acquisition time, many NMR logging tools often acquire short echo trains with a short repeat delay time that are stacked many times. The limitation of using an insufficient repeat delay time is overcome by either assuming a fixed T_1/T_2 ratio of 1.65 or by selecting an optimized T_1/T_2 ratio between 1 and 3 using the high-precision regular echo train and ample repeat delay for the best fit [20]. Such a practice often works but runs into problem when the earth formation contains a large amount of paramagnetic impurities. Recently, Sun et al. [21] introduced a functional relationship between the T_1/T_2 ratio and T_2, as follows:

$$\mathrm{Log}(T_1/T_2) = a - b \cdot \log(T_2) \tag{7.4}$$

where a and b are two adjustable parameters.

Figure 7.2 demonstrates an application of this function when estimating the formation porosity from the log data with Schlumberger's CMR logging tool. As displayed in track 3, given the optimal T_1/T_2 ratio, the porosity values derived from NMR (CPHT, pink solid line) have a good agreement with the values obtained from the neutron/density cross plot (PHTI, red solid line). The NMR data also offers a reasonable estimate of the zone's pore-size distribution from the T_2 distribution (track 6) and the water/oil saturation at a different depth when a proper T_2 cutoff (red vertical line) is applied.

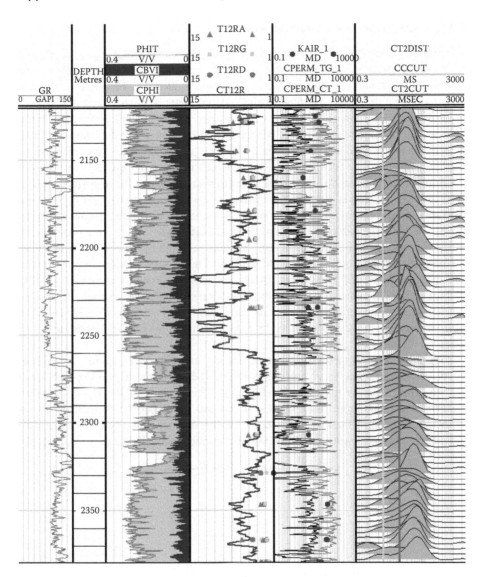

FIGURE 7.2 CMR log was reprocessed with a maximum possible T_1/T_2 ratio of 20. Track 1 shows the gamma ray log; track 2 shows the depth log; track 3 shows the comparison between the NMR-derived porosity (CPHI) and PHIT derived from neutron/density cross plot; track 4 is the T_1/T_2 ratio log (solid line) derived from CMR T_2 inversion and T_1/T_2 ratio (dots) from core measurements; track 5 shows NMR-derived permeability based on Coates and T_{2G} equations; and track 6 shows NMR T_2 distribution. (From B. Sun et al., The impact of T_1/T_2 ratio on porosity estimation. *SPWLA 49th Annual Logging Symposium, Edinburgh, Scotland*, May 25–28, 2008. Copyright 2008, Society of Petroleum Engineers Inc. Reproduced with permission of SPE. Further reproduction prohibited without permission.)

7.4.2 IRREDUCIBLE WATER DETERMINATION

The porosity and pore-size information from NMR measurements can also be used to determine the amount of producible fluid and to estimate permeability values. The amount of producible fluid (represented by the free fluid index, FFI) is based on the assumption that the producible fluid resides in large pores whereas bound fluid resides in small pores. The unit of FFI is p.u., where 1 p.u. = 1% of the bulk rock volume. The part of fluid in the formation that is not producible is termed "irreducible water," and reported as the bulk volume irreducible (BVI). The irreducible water saturation can be loosely defined as the water saturation below which increase in capillary pressure will not result in a significant decrease in the saturation value. One of the important applications of NMR logging is estimating the irreducible water saturation, S_{wirr} [3]. Since the relaxation time is proportional to pore size, a T_2 value can be selected below which the corresponding fluids are considered as irreducible and above which the corresponding fluids are expected to be producible. This T_2 value is called the T_2 cutoff. These values are best determined through laboratory measurements of rock samples from each field and are typically around 33 ms for sandstones and around 100 ms for carbonates [22,23].

Figure 7.3 shows a cross plot of the FFI determined by centrifuging a suite of sandstone plugs at an equivalent capillary pressure of 100 psi. This plot was obtained by comparing with the FFI obtained by the cutoff at 33 ms on the T_2 measurements made on the same plug samples.

Figure 7.4 demonstrates a similar cross plot for carbonate rocks where a T_2 cutoff of 92 ms was used. In practice, the most appropriate T_2 cutoff for sandstones or carbonates can vary from reservoir to reservoir and usually fall in certain ranges. The T_2 cutoff for sandstones can vary from 20 to 50 ms, whereas that for carbonates is usually from 90 to 150 ms. In general, the T_2 cutoff values for carbonates are typically much larger than those for sandstones, owing to the much smaller surface relaxivity for the carbonates [2].

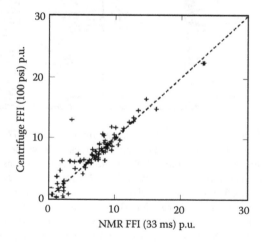

FIGURE 7.3 Centrifuge FFI of a suite of sandstone plugs at 100 psi versus the NMR FFI calculated using a T_2 cutoff at 33 ms. The dashed line is 1:1. (From C. Straley et al., *Log Analyst*, 1997, 38:84–94. Presented at the 1994 International Symposium, Society of Core Analysts, SCA9404, Stavanger, Norway.)

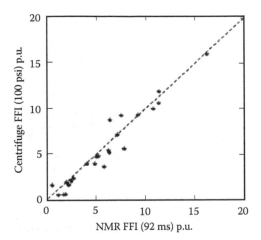

FIGURE 7.4 Centrifuge FFI of a suite of carbonate plugs at 100 psi versus the NMR FFI calculated using a T_2 cutoff at 92 ms. The dashed line is 1:1. (From C. Straley et al., *Log Analyst*, 1997, 38:84–94. Presented at the 1994 International Symposium, Society of Core Analysts, SCA9404, Stavanger, Norway.)

7.4.3 PERMEABILITY PREDICTION

Permeability is the rock property that indicates the ability of fluids or gases to flow through. Because the relaxation time is proportional to the pore size, or more specifically the ratio of pore volume to surface area, it is also related to permeability. The permeability of an earth formation is extremely important in predicting the fluid producibility, the subsequent development plan, and the management of the reservoir over the life span of its active production. No well logging technique, other than NMR, can provide a systematic estimation of permeability [2]. The NMR estimation of permeability is based on a combination of experimental and theoretical models. It has been reported in many studies that the permeability in reservoirs is related to the irreducible water saturation (S_{wirr}). The corresponding quantities that can be estimated from NMR measurements are the volume of free fluid (FFI, $(1 - S_{wirr})\Phi$) and the volume of irreducible water (BVI, $S_{wirr}\Phi$). Figure 7.5 illustrates how the T_2 distribution is composed of movable (FFI) and immovable (BVI) components. One of the most widely used empirical correlations based on the irreducible water saturation, S_{wirr}, for estimating permeability is the Coates equation [24]:

$$\kappa = \left[\left(\frac{\phi}{10} \right)^2 \frac{FFI}{BVI} \right]^2 \tag{7.5}$$

where κ is the permeability in millidarcy, ϕ is the porosity in p.u., and FFI and BVI are in p.u.

An example of using the Coates equation to predict reservoir permeability in NMR well logging is shown in track 5 of Figure 7.2, where the permeability calculated using Equation 7.5 is demonstrated with a black solid line.

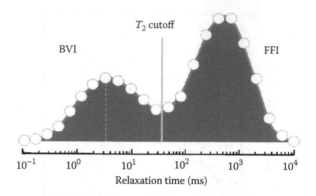

FIGURE 7.5 Coates (or free fluid) model is the simplest and the most commonly used model; it uses FFI/BVI ratio to determine permeability. (From G.R. Coates and J.L. Dumanoir, *Log Analyst*, 1974, 15(1):17–29.)

7.4.4 ESTIMATING VISCOSITY

The viscosity of hydrocarbon in the formation is one of the key properties for the evaluation and development of petroleum reservoirs, and the NMR log is the only log thus far that provides information on oil viscosity [3]. The empirical relationship developed in the laboratory is often used to predict the viscosity of the crude oils in the earth formation. Once the T_2 signal for the crude oil is identified and separated from the water signal, its logarithmic mean value is employed for the estimation of viscosity.

The correlation between T_2 relaxation times and viscosity for dead crude oils (containing dissolved oxygen) reported by Morriss et al. [25] is one of the most commonly used empirical correlations used in the industry (Equation 7.6):

$$T_{2,\text{LM}} = 0.00403 \frac{T_\text{K}}{\eta} \tag{7.6}$$

where $T_{2,\text{LM}}$ is in seconds and refers to the logarithmic mean of oil T_2, T_K is the absolute temperature in Kelvin, and η is the viscosity in centipoise (cP). As shown in Figure 7.6, the Morriss correlation works very well for light- to medium-viscosity oils (10–1000 cP) but fails for the heavy oils with a viscosity of >1000 cP. Later work by Zhang et al. [26] and Lo et al. [27] on deoxygenated pure alkane and alkane mixtures also established a linear relationship between T_2 and viscosity on the log–log scale.

In studies on heavy crude oils using low-field NMR spectrometers, the apparent T_2 values were found to depend on the echo spacing (TE) applied in the measurements [5,6,15], where echo spacing is the time between successive echoes in a CPMG echo train during an NMR measurement. It is also the time to the first echo and is therefore an important parameter in defining the fastest relaxation time that can be measured. As shown in Figure 7.7, the T_2 values are highest for the longest TE because of the loss of short components as TE is increased. LaTorraca et al. [5]

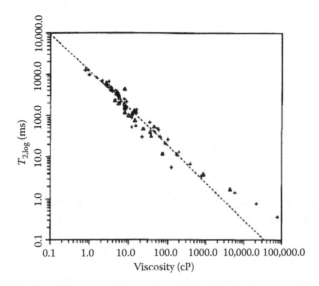

FIGURE 7.6 Cross plot of logarithmic mean T_2 versus viscosity of bulk crude oils (various symbols). The dashed line is calculated using Equation 7.6. (From C.E. Morriss et al., Hydrocarbon saturation and viscosity estimation from NMR logging in the Belridge diatomite. Paper C. Presented at SPWLA 35th Annual Logging Symposium, June 19–22, Tulsa, OK: SPWLA, 1994.)

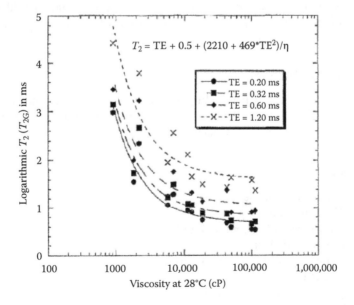

FIGURE 7.7 Logarithmic mean of T_2 for different heavy oils obtained with varying TE versus viscosity using a 2-MHz NMR spectrometer. (From G.A. LaTorraca et al., Heavy oil viscosity determination using NMR logs. Paper PPP. Presented at SPWLA 40th Annual Logging Symposium, May 30–June 3, Oslo, Norway: SPWLA, 1999.)

also established a formula to correct this echo spacing dependence in regular T_2 measurement:

$$\eta = \left(\frac{2210 + 469 \cdot TE^2}{T_{2,LM} - (TE + 0.5)} \right) \cdot \frac{T_K}{298} \qquad (7.7)$$

where $T_{2,LM}$ and TE are in milliseconds and η is in centipoise.

More recently, Sun et al. [15] and Yang and Hirasaki [6] developed new correlations to overcome the echo spacing restriction during heavy oil measurements.

7.5   OUTLOOK ON THE FUTURE OF NMR TECHNOLOGY FOR OIL EXPLORATION AND CHARACTERIZATION

NMR logging tools and data analyses have come a long way since their first introduction in the early 1990s. The superiority of NMR logging tools at providing valuable petrophysical information makes these tools stand out among all logging devices. Currently, NMR logging tools are extensively used for determining reservoir porosities and permeabilities [28], and for real-time analysis of bottom-hole sampling of reservoir fluids [29,30]. The introduction of multifrequency multiprobe-zone logging tools makes it possible to perform both fluid characterization and thin bed analysis. However, NMR logging is expensive and requires slower logging speed compared with traditional logging tools. Future tool development shall focus on a deeper probing zone, improved signal-to-noise ratio, and a wider range of field gradients to accommodate a larger range of diffusion constants for heavier oils and logging technologies while drilling.

ACKNOWLEDGMENTS

We thank Dr. Scott Seltzer, Chevron Energy Technology Company, and the reviewers for their valuable suggestions. We also thank Chevron Energy Technology Company for the permission to publish this chapter.

GLOSSARY

BVI: bulk volume irreducible
CPHI: NMR-derived porosity
CPMG: Carr–Purcell–Meiboom–Gill pulse sequence
D: self-diffusion coefficient
FET: fluid typing by editing T_2 distribution
FFI: free fluid index
GIFT: global inversion of fluid typing
κ: permeability
PHIT: porosity derived from neutron-density cross plot
S_{wirr}: irreducible water saturation
T_1: longitudinal relaxation time

T_{1B}: longitudinal relaxation time for the bulk fluid
T_{1S}: longitudinal relaxation time caused by the surface
T_2: transverse relaxation time
T_{2B}: transverse relaxation time for the bulk fluid
T_{2D}: transverse relaxation time caused by the diffusion effect
$T_{2,LM}$: logarithmic average of the T_2 relaxation time
T_{2S}: transverse relaxation time caused by the surface
TE: time between successive echoes in a CPMG echo train
η: viscosity
Φ: porosity of formation

REFERENCES

1. Cowan, B. *Nuclear Magnetic Resonance and Relaxation.* Cambridge, UK: Cambridge University Press, 1997.
2. Coates, G.R., L. Xiao, and M.G. Prammer. *NMR Logging: Principles of Applications.* Houston, TX: Halliburton Energy Service, 1999.
3. Dunn, K.J., D.J. Bergman, and G.A. LaTorraca. *Nuclear Magnetic Resonance Petrophysical and Logging Applications.* New York: Pergamon, 2002.
4. Bryan, J., A. Mai, F. Hum, and A. Kantzas. Oil- and water-content measurements in bitumen ore and froth samples using low-field NMR. *SPE Reservoir Eval. Eng.* 2006, 9:654–663.
5. LaTorraca, G.A., S.W. Stonard, P.R. Webber, R.M. Carlson, and K.J. Dunn. Heavy oil viscosity determination using NMR logs. *40th SPWLA Annual Logging Symposium.* Oslo, Norway: SPWLA, 1999. Paper PPP.
6. Yang, Z., and G.J. Hirasaki. NMR measurement of bitumen at different temperatures. *J. Magn. Reson.* 2008, 192:280–293.
7. Yang, Z., G.J. Hirasaki, M. Appel, and D.A. Reed. Viscosity evaluation for NMR well logging of live heavy oils. *Petrophysics* 2012, 53:22–37.
8. Brownstein, K.R., and C.E. Tarr. Importance of classical diffusion in NMR studies of water in biological cells. *Phys. Rev. A* 1979, 19:2446.
9. Brown, R.J., and I. Fatt. Measurement of fractional wettability of oil field rocks by the nuclear magnetic relaxation method. *Trans. AIME* 1956, 207:262–264.
10. Carr, H.Y., and E.M. Purcell. Effects of diffusion on free precession in nuclear magnetic resonance experiments. *Phys. Rev.* 1954, 94:630–638.
11. Meiboom, S., and D. Gill. Modified spin-echo method for measuring nuclear relaxation times. *Rev. Sci. Instrum.* 1958, 29:688–691.
12. Sun, B., and K.-J. Dunn. Methods and limitations of NMR data inversion for fluid typing. *J. Magn. Reson.* 2004, 169:118–128.
13. Hürlimann, M.D., and L. Venkataramanan. Quantitative measurement of two-dimensional distribution functions of diffusion and relaxation in grossly inhomogeneous fields. *J. Magn. Reson.* 2002, 157:31–42.
14. Sun, B., and K.-J. Dunn. A global inversion method for multi-dimensional NMR logging. *J. Magn. Reson.* 2005, 172:152–160.
15. Sun, B., K.-J. Dunn, G.A. LaTorraca, C. Liu, and G. Menard. Apparent hydrogen index and its correlation with heavy oil viscosity. *SPWLA 48th Annual Logging Symposium Transactions.* Austin, TX: SPWLA, 2007. Paper R.
16. Song, Y.Q., L. Venkataramanan, M.D. Hürlimann, M. Flaum, P. Frulla, and C. Straley. T_1-T_2 correlation spectra obtained using a fast two dimensional LaPlace inversion. *J. Magn. Reson.* 2002, 154:261–268.

17. Sun, B., and K.-J. Dunn. Core analysis with two dimensional NMR. *Proceedings from the 2002 International Symposium of the Society of Core Analysts.* Monterey, California, September 22–25, 2002.
18. Hearst, J.R., and P.H. Nelson. *Well Logging for Physical Properties.* New York: McGraw-Hill Book Company, 1985.
19. Vinegar, H.J. X-ray CT and NMR imaging of rocks. *J. Pet. Technol.* 1986, 38:115–117.
20. Kleinberg, K.L., S.A. Farooqui, and M.A. Horsfield. T_1/T_2 ratio and frequency dependence of NMR relaxation in porous sedimentary rocks. *J. Colloid Interface Sci.* 1993, 158(1):195–198.
21. Sun, B., M. Skalinski, J. Brantjes, G.A. LaTorraca, G. Menard, and K.-J. Dunn. The impact of T_1/T_2 ratio on porosity estimation. *SPWLA 49th Annual Logging Symposium.* Edinburgh, Scotland, May 25–28, 2008.
22. Dunn, K.J., G.A. LaTorraca, J.L. Warner, and D.J. Bergman. On the calculation and interpretation of NMR relaxation time distribution, *1994 SPE Annual Technical Conference and Exhibition Proceedings.* New Orleans, LA: Society of Petroleum Engineers, 1994, pp. 45–54.
23. Straley, C., D. Rossini, A. Vinegar, P.N. Tutunjian, and C.E. Morriss. Core analysis by low-field NMR. *Log Analyst* 1997, 38:84–94.
24. Coates, G.R., and J.L. Dumanoir. A new approach to improved log-derived permeability. *Log Analyst* 1974, 15(1):17–29.
25. Morriss, C.E., R. Freedman, C. Straley, M. Johnston, H.J. Vinegar, and P.N. Tutunjian. Hydrocarbon saturation and viscosity estimation from NMR logging in the Belridge diatomite. *35th SPWLA Annual Logging Symposium Transactions.* Tulsa, OK: SPWLA, 1994. Paper C.
26. Zhang, Q., S.-W. Lo, C.C. Huang, G.J. Hirasaki, R. Kobayashi, and W.V. House. Some exceptions to default NMR rock and fluid properties. *SPWLA 39th Annual Logging Symposium.* Keystone, CO: SPWLA, 1998. Paper FF.
27. Lo, S.W., G.J. Hirasaki, W.V. House, and R. Kobayashi. Mixing rules and correlations of NMR relaxation time with viscosity, diffusivity and gas/oil ratio of methane/hydrocarbon mixtures. *SPE J.* 2002, 7:24–34.
28. Prammer, M.G. NMR pore size distributions and permeability at the well site. *1994 SPE Annual Technical Conference and Exhibition Proceedings.* New Orleans: SPE, 1994, pp. 55–64.
29. Bouton, J., M.G. Prammer, P. Masak, and S. Menger. Assessment of sample contamination by down-hole NMR fluid analysis. *2001 SPE Annual Technical Conference and Exhibition Proceedings.* New Orleans, LA: SPE, 2001.
30. Masak, P.C., J. Bouton, M.G., Prammer, S. Menger, E. Drack, B. Sun., K.-J. Dunn, and M. Sullivan. Field test results and applications of the down-hole magnetic resonance fluid analyzer. *43rd SPWLA Annual Logging Symposium Transactions.* Oiso, Japan: SPWLA, 2002. Paper 2002-GGG.

8 Nuclear Magnetic Resonance Upstream Applications
Crude Oil Characterization, Water–Oil Interface Behavior, and Porous Media

Teresa E. Lehmann and Vladimir Alvarado

CONTENTS

CONTEXT

- Further applications of NMR technology for the study of water–oil interfacial problems, and effect of petroleum and heavy fractions in porous media
- Potential applications of NMR imaging techniques, and in polymer-flooding problems

ABSTRACT

In this chapter, a summary of the characterization of crude oils and their derivatives through nuclear magnetic resonance (NMR) spectroscopy is presented. The use of this technique for the study of asphaltenes, water–oil interfacial problems, and oil in porous media is also discussed. Additionally, several new and potential applications of NMR are introduced for an outlook of emerging techniques in the oil and gas industry. These include the use of new data analysis methods, use of imaging techniques, and use of NMR in polymer-flooding problems.

8.1 INTRODUCTION

Nuclear magnetic resonance (NMR) is a nondestructive technique that can be used to characterize a variety of systems and oil derivatives such as preserved reservoir rock samples, micro- and macro-emulsions, as well as crude oils and their fractions, e.g., asphaltenes, resins, and organic acids. Different functional groups in a molecule resonate in unique, well-resolved spectral regions, and this fact has established NMR as a highly discriminative and quantitatively reliable analytical technique. NMR has been widely employed to characterize oil samples from different angles, including molecular structures, chemical composition, kinetic behavior, and oil-in-water dispersion stability. While analytical techniques, and NMR in particular, have been heavily used in downstream applications in the oil and gas industry, namely in refining and petrochemistry, the transfer and adaptation of some of these analytical techniques to upstream applications have been increasingly taking place in more recent times. High-field NMR has been the domain of analytical chemistry when it relates to these types of applications; however, low-field or time-domain NMR has fulfilled many of the needs in the upstream world. Relaxometry has become a mature and frequently used low-field NMR alternative, particularly with regard to characterization of porous media and heavy crude oils. Some of the most relevant work employing NMR in these areas is discussed in this chapter.

The chapter is organized as follows. A summary of characterization of crude oils and their derivatives is presented. We then proceed to elaborate on a solubility class of compounds that are of importance throughout the value chain in the oil and gas industry, namely asphaltenes. This solubility class of compounds is broadly defined as the oil fraction insoluble in low molecular weight normal alkanes such as n-pentane and n-heptane, but soluble in aromatics such as benzene and toluene. This solubility class plays an important role in formation damage problems (porous medium permeability reduction) as well as in production problems such as the formation of stable, viscous water–oil emulsions [1].

Although emphasis has been placed on asphaltenes, among solubility classifications in crude oil, other fractions also participate in interfacial mechanisms and the stability of asphaltenes. Resins have been argued to intervene in solvation of the asphaltene fractions, and hence are intrinsic participants in interfacial mechanisms between water–crude oil, fluids, and rock. Along this vein, we follow with a discussion and experimental results on the role of naphthenic acids (NAs) in water–oil interfacial problems and their characterization via solution NMR [2].

Porous media have been studied using imaging techniques as well as time-domain or low-field NMR. We review principles and applications regarding porous medium characterization. Finally, several new and potential applications of NMR are introduced for an outlook of emerging techniques in the oil and gas industry.

8.2 CHARACTERIZATION OF CRUDE OILS AND THEIR DERIVATIVES

NMR has been widely applied in the oil industry for the characterization of different compounds in crude oils. The solid content (wax precipitation) of crude oils has been determined as a function of temperature for North Sea crude oils [3]. The method involved the measurement of the amplitude of the NMR signal 10 and 70 μs after the application of a radiofrequency pulse. Considering that the NMR signal decays after 10 and 70 μs are proportional to the number of protons in the solid and liquid phases, respectively, it was possible for these authors to determine the solid-state phase content that correlated well with values measured through acetone precipitation.

NMR has also been used to determine the distribution of hydrogen and carbon in heavy oil (aromaticity). The results of these experiments were used to investigate the molecular structure of some weathered oils from Kuwait's environment, to clarify the direction and the speed of weathering [4]. Comparison of the average molecular weights of six oils extracted from sediment in different regions of Kuwait indicated that oils in soils are degraded by condensation, aromatization, cyclization, and oxidation reactions. On the other hand, oils in sediment are degraded by the formation of naphthenic rings and the decrease of aliphatic chains.

The amount of information on hydrocarbons with ethyl, *n*-propyl, *n*-butyl, and other substituents on the carbon chains (T-branched hydrocarbons) in topped crude oils and in bitumens of dispersed organic matter is extremely small. An NMR-based method for the identification of these compounds was developed by Smirnov and Vanyukova [5] through the use of ^{13}C-NMR spectroscopy. The use of this method could make the determination of these compounds be a routine procedure.

In another application, magic angle spinning/cross-polarization (MAS/CP) NMR has been applied to frozen fossil fuels for the analysis of carbon center types [6]. This strategy was useful at providing compositional information on single petroleum components, petroleum distillates, and lube oil stocks and their refined products. This type of study provides a relatively simple way to compare and correlate crudes according to their gross compositional features.

Multinuclear NMR is widely employed in the oil industry. In this area, ^{1}H-, ^{13}C-, and J-coupling NMR experiments have been used to yield information on the relative percentages of paraffinic, naphthenic, and aromatic structures in nitrogen-free vacuum gas–oil distillate from a Kuwaiti crude oil [7]. This analysis is important for refinery product quality assurance in geological studies, and in crude oil assessment. In previous work in this area, ^{1}H- and ^{13}C-NMR spectroscopies were used to elucidate the chemical structures of heavy oils from petroleum, vacuum residues, asphaltenes, and liquid coal [8–13]. ^{13}C-NMR data have also been used to extend the potentialities of comparative analysis of petroleum composition usually performed through gas–liquid chromatography and gas chromatography–mass spectrometry techniques [14].

NMR-derived parameters and structures have also become valuable in upstream applications. For example, the aromaticity index determined from NMR is an indication of the growth of polyaromatic systems. Measurements of this index are usually performed on heavy petroleum residues from different refineries, which can be used as raw material in the production of various carbon products such as isotropic and anisotropic pitches and different types of coke [15,16]. It can also be used to determine wax precipitation at low temperatures [17].

The increasing need for crude oil has made crude blending refining increasingly popular. However, blending refining can cause a change in the properties of crude, fractions, and residua that needs to be properly assessed. NMR and typical analyses have been performed on crude mixtures to estimate the mean structure of blended crudes at the optimal transition state with excellent results [18].

Heavy fuel oil is a complex, highly heterogeneous mixture of saturated, unsaturated, and aromatic hydrocarbons, containing additional elements such as sulfur, vanadium, and nitrogen. The molecules contained within this mixture range from simple hydrocarbons to complex macromolecular structures. Because of the high level of complexity and the large number of individual, structurally different components, the complete qualitative and quantitative characterization of each constituent in the oil is challenging. NMR spectroscopy, in combination with multivariate data analysis, has been used for quick and accurate extraction of parameters pertaining to the physical and chemical properties of these complex suspensions. Good prediction models were obtained for a large number of characterization parameters, including the calculated aromaticity index, density, gross and net calorific values, and water and sulfur contents, as well as micro-carbon residue [19].

Crude oil is a mixture of a very large number of different hydrocarbons. Each petroleum variety has a unique mix of molecules that define its physical and chemical properties and ultimately its behavior during refining. Currently, refiners have to cope with an increasing demand of high-quality distillates of tighter specifications, while facing the need to process increasingly heavier and poorer quality crudes. NMR spectroscopy in combination with partial least squares (PLS) regression models was successfully used to predict the properties of crude oil samples. The results of this investigation indicate that this method can be applied with success to modern refining process requirements [20].

The changes in bitumen structure of fresh, aged, and doped recycled bitumens were investigated using rheology and NMR [21]. An inverse Laplace transform of the NMR spin-echo decay (T_2) was applied to draw a map of the microaggregates inside bitumen at different temperatures. The results of this investigation indicate that the rejuvenating additives have little influence on the solid–liquid transition. However, they help restructure the aged bitumen, thus facilitating workability at a lower temperature.

NMR spectroscopy has also influenced heavy crude oil characterization. The worldwide reduction in economically recoverable conventional petroleum reserves has led to an increase in exploration and production activity in heavier crude oils. Diversified Canadian crude oils, including oil sands, bitumens plus heavy and conventional oils, were characterized through NMR spectroscopy. The results of this investigation indicated that the bitumens comprise less saturates but more resins and asphaltenes than any of the other heavy oils tested. Conversely, the conventional crude is associated with the highest saturates content and the least amount of resins and asphaltenes [22]. A thorough knowledge of the molecular structure and behavior of the source petroleum is needed to optimize plant-operating conditions and assess their impact on the environment.

NMR spectroscopy is the technique of choice for the study of biological systems, and this particular aspect has also shown to be useful in upstream applications. NMR was used to characterize biosurfactant production by various strains of *Bacillus subtilis* isolated from Brazilian crude oils [23]. The biosurfactant mixtures identified were able to decrease the interfacial tension of Arabian Light crude oil–water system more efficiently than chemical surfactants, and showed better results in oil recovery, thus suggesting their interest for use in microbial-enhanced oil recovery processes.

8.3 ASPHALTENES

Asphaltenes form a compound class that deserves special attention because of the problems associated with them in oil production and refining processes. For this reason, the understanding of the behavior and composition of asphaltenes has been actively pursued in the oil industry, and NMR spectroscopy has played an important role in this area. NMR has been used in the elemental analysis of many asphaltenes [24], to determine their average hydrocarbon structure [25–28], and for the structural characterization of asphaltenes of different origins [29]. Gated spin echo (GASPE) ^{13}C-NMR spectroscopy was used to determine the presence of CH groups in the asphaltene aliphatic chains, and correlate the CH_3/CH ratio with aromaticity. The degree of condensation and/or substitution of aromatic rings were evaluated by applying ^{13}C solid-state NMR. These studies indicated that asphaltenes with different aromaticity seem to be similar in their aromatic ring condensation and/or substitution degrees. In another application, ^{13}C- and ^{1}H-NMR spectroscopy showed that isomerization, internal cross-linking, and dehydrogenation of Arabian asphalt were the main chemical reactions of hydrocarbon groups following aging [30]. Additionally, the mobility of asphaltenes derived from different crude oils was determined through ^{13}C-NMR and MAS/CP experiments, in an attempt to establish the structural differences between them [31].

Two-dimensional NMR spectroscopy has also provided important insight into the study of ashphaltenes. Diffusion-ordered two-dimensional NMR spectroscopy (DOESY) was used to yield structural and dynamic information (molecular size and aggregation state) in different types of oil fractions ranging from kerosene cuts to fractions that have an initial boiling point above 560°C [32,33]. Asphaltenes derived from the oils examined were characterized as "continental," based on the fact that the aromatic species found were highly substituted and the polycyclic aromatic hydrocarbons were connected to long alkyl chains [33]. Buzurgan asphaltene samples have also been analyzed through DOESY spectroscopy [34]. Parameters such as molecular weight and particle size were derived for the asphaltenes investigated in this study. The molecular dynamics of asphaltenes was explored by studying the influence of the asphaltene concentration on their mobility in solution. It was determined that in the dilute regime, the diffusion coefficients remain constant as only solute–solvent interactions limit the translational self-diffusion, whereas more intermolecular interactions start to be involved above 0.25 wt.%. From this observation, the onset of the aggregation process was defined at 0.25 wt.%.

The possibility of using NMR to study solid compounds has provided yet another avenue for the study of asphaltenes. The structural changes in Kuwait asphaltenes occurring during pyrolysis have been studied by solid-state ^{13}C-NMR [35]. The single pulse excitation/MAS technique was applied to determine the ratio of aromatic to aliphatic carbon, and the CP/IP/MAS technique was used to estimate the ratio of tertiary to quaternary aromatic carbon. The most obvious effect of thermal cracking of the asphaltenes was determined to be the remarkable increase of the ratio of aromatic to aliphatic carbon in the samples.

8.4 ROLE OF NAPHTHENIC ACIDS IN WATER–OIL INTERFACIAL PROBLEMS

Emulsions are ubiquitous to oil production operations and remain a challenge in conventional and enhanced oil recovery operations [1]. Emulsions are generally defined as a dispersion of one liquid into another liquid phase, but can be more complex by having an immiscible phase, e.g., water, dispersed in a dispersed oleic phase, which in turn is dispersed in a continuous aqueous phase. These are the so-called multiple emulsions as opposed to simpler oil-in-water or water-in-oil emulsions. In general, water and crude oil are produced in the form of macroemulsions. This type of dispersions is thermodynamically unstable; however, phase separation can be suppressed by the existence of kinetic barriers to coalescence. Organic crude oil fractions such as asphaltenes, resins, and organic acids, as well as waxes and inorganic particles (clays and other fines) can contribute to the dynamic stability of macroemulsions. Asphaltene layers in water–oil interfaces have been identified as the main reason for the stability of water-in-oil emulsions [36]. Emulsion breakers are necessary to produce market-quality products. On the other hand, and although emulsions are often perceived as being detrimental in production operations, experimental and field evidence exists to show that benefits might be derived in terms of enhanced oil recovery [37]. NMR is a noninvasive, nondestructive technique that can be used for high concentrations of the dispersed phase. Since this method does not rely on the transparency of the continuous phase, NMR can produce size distributions in opaque systems such as water-in-crude oil emulsions.

The dynamic stability of emulsions has been studied through a variety of techniques (see Reference 1, for details). It turns out that the proximity to interfaces affects relaxation mechanisms, and therefore this enables proxies for drop size. NMR sensitivity depends on the strength of the external magnetic field and the nuclear moment, which in turn is a function of the magnetogyric ratio. The structure of organic molecules or the diffusion coefficient can be determined [38,39].

When applied to emulsions, NMR techniques rely on signal attenuation associated to the random motion of molecules when two magnetic field gradients are imposed on the system, as the NMR signal intensity is directly proportional to the number of nuclei that generate it. The method developed this way is called NMR pulsed field gradient–spin echo (PFG–SE) and was first demonstrated by Tanner and Stejskal [40] to measure the free unrestricted self-diffusion of molecules. The signal attenuation for a single diffusing molecule is described by

$$\frac{S}{S_0} = \exp\left(-D(\gamma g \delta)^2 \left(\Delta - \frac{\delta}{3}\right)\right)$$

(8.1)

where D is the self-diffusion coefficient of the relevant molecules $\left(\dfrac{cm^2}{s}\right)$ and Δ, δ, and g are the temporal duration between the application of the two magnetic field gradients (s), the duration of a single magnetic field gradient (s), and the strength of the magnetic field gradient $\left(\dfrac{T}{cm}\right)$, respectively. γ ($T^{-1}S^{-1}$) is the gyromagnetic ratio of the nucleus being studied [41]. Equation 8.1 can be used in unrestricted diffusion cases, namely when no barriers for diffusion exist. Water-in-oil emulsions present a case in which molecular diffusion is restricted to the spherical cavity represented by water droplets. By varying the duration of the applied magnetic field gradient, it is possible to observe the effect of the cavity wall on diffusion and consequently cavity dimensions [42]. Two methods are usually followed to estimate the drop size [41]:

1. *Gaussian phase distribution model (GPD).* This method was first formulated by Murday and Cotts [42] and assumes that diffusion following the gradient pulses is Gaussian phase distributed. The signal attenuation for the spherical cavity has a complex relationship with the radius of the sphere (droplet radius).
2. *Short gradient pulse (SGP) model.* This model assumes that the duration of the applied magnetic field gradient is zero. The equations describe magnetic responses; the radius of the emulsion droplet is the only free parameter and all other parameters are determined in the experiment. Since emulsions are generally polydispersed, echo signals coming from different droplet sizes are superimposed. This can be solved by assuming the distribution function for droplet size. The measured signal in this case can be defined by the following integral:

$$\frac{S}{S_0} = \frac{\displaystyle\int_0^\infty a^3 P(a) R(g,\delta,a,\Delta)\,\mathrm{d}a}{\displaystyle\int_0^\infty a^3 P(a)\,\mathrm{d}a} \qquad (8.2)$$

where $R(g,\delta,a,\Delta)$ is the signal attenuation function and $P(a)$ is the droplet size distribution function. Typically, a log-normal distribution function is used to describe the size distribution of the emulsions [40]. This restriction results in the loss of the actual size distribution of the emulsion. Therefore, some alternative methods have been developed that do not assume any predetermined distribution function for the system and yield more general information about droplet size distributions [41–44]. In another application, the attenuation of some NMR signals can be due to restricted molecular diffusion within the droplets or to the diffusion of droplets by themselves. This was studied by Garasanin et al. [45].

Honorato et al. [46] showed that a combination of low- and high-field NMR techniques can be used to study changes in heavy crude oil upon treatment with plasma. They showed the separation of effects caused on the water and oil phases, not unlike the effects on emulsion systems. A Carr–Purcell–Meiboom–Gill (CPMG) pulse sequence was employed to measure T_2. As is typical in this type of applications, distributions of relaxation times were obtained by deconvolution methods (e.g., inverse Laplace transform) to correlate T_2 with physical properties of the fluids, e.g., temperature-dependent viscosity. High-field NMR results allowed the authors to examine chemical shifts, which would have indicated chemical changes upon application of plasma, which were not observed in their case. This interesting combination of NMR techniques provided insights about viscosity-reduction mechanisms, which were mostly associated with demulsification upon plasma treatment. PFG–SE NMR [47] spectroscopy and the low-field NMR technique CPMG [48–50] have also been used to determine the droplet size in oil-in-water emulsions to determine emulsion stability.

Moradi et al. [2] used an experimental protocol called a partitioning test in which crude oil and an aqueous phase are placed in contact to partition aqueous soluble species in the crude oil into the aqueous phase. In this simple experiment, the crude oil–water interface is allowed to exchange components in a similar fashion as water and crude oil in emulsion systems. ^{1}H-1D NMR was used to study changes in the overall NA content in water during partitioning experiments, without distinction of specific species in the family of organic acids leached off the crude oil phase. To develop a semiquantitative method, NMR spectra were calibrated to correct for aqueous-phase electrical conductivity by using spectra obtained with a standard. This enables the determination of relative change in concentration of NAs as a function of aqueous-phase ionic strength.

In the experiments described in this work, a crude oil from the state of Wyoming labeled GF (19°API) was used in all assays (oil density is 0.91 g/mL at 22°C). The C7-asphaltene content is 9%, as received. Deuterium oxide (98% D) and decane (from Sigma Aldrich) were used as received. A commercial blend of NAs, and p-aminobenzoic acid (PABA) were purchased from Acros Organic/Fisher Scientific. Analytical-grade $CaCl_2$ was purchased from Fisher Scientific. Aqueous solutions were prepared by dissolving analytical-grade $CaCl_2$ in dionized water. The ionic strength of the aqueous phase was 0.6724 M, which was labeled 100% salinity for simplicity. Salinity was modified by diluting with deionized water to obtain 10%, 1%, and 0.1% ionic strength relative to the parent aqueous phase. The conductivity was determined using an Accumet Basic AB30 conductivity meter at room temperature. The meter was calibrated before any measurement using standard solutions of KCl with conductivity values of 45, 1500, and 45,000 ms/cm.

Partition tests, in which the crude oil and a selected aqueous phase at a volume ratio of 1:1 are placed in contact at ambient temperature for a predetermined time, were conducted to allow exchange of components between the oleic and aqueous phases. In this experiment, water of known salinity (0.1%, 1%, 10%, and 100%) is added to a beaker first and then the selected crude oil is carefully poured on top to minimize mixing with water. A hollow tube is placed before pouring the crude oil to allow collection of water aliquots as well as to measure conductivity and pH, if needed. The crude oil and aqueous phases are left in contact for 1, 2, 5, and 7 days. After aging for the aforementioned period, a J-shaped needle is used to collect 1 mL of the aqueous phase in proximity to the water–oil interface through the hollow tube. Aliquots of the aqueous phase were collected using a J-shaped needle through the hollow tube at desired times.

Once a partitioning experiment was set up, NMR samples were prepared by dissolving 540-mL aliquots of the partitioned brine with 60 mL D_2O to afford 90% H_2O/10% D_2O mixtures. NMR experiments were performed at 600 MHz on a Bruker Avance III 600 instrument. All chemical shifts were referenced to H_2O as the internal standard. All spectra were collected at 25°C with widths of 16 ppm. A 11.5-s 90° pulse was calibrated for the samples. 1D ^{1}H-NMR spectra were collected with WATERGATE incorporated into the pulse sequence to achieve solvent suppression, and 1024 data points and 512 scans were collected. Signal-to-noise ratio was improved by applying a line-broadening factor of 2 Hz to the free induction decay before Fourier transformation. PABA was used as a standard to calibrate the intensity of NMR spectra with respect to aqueous-phase electrical conductivity. PABA was dissolved in $CaCl_2$ brines with salinity values of 0.05%, 0.1%, 1%, 5%, 10%, and 100% to prepare solutions to a concentration of 0.2 mM. The 1D ^{1}H-NMR spectra of these samples were collected under the same conditions listed above for the partitioned brine samples. Selected peak intensities and their associated signal integral versus the conductivity of each solution were plotted. An exponential fit through the experimental data served as the calibration curve. This procedure enabled obtaining intensity attenuation as a function of conductivity. This factor was used to correct the intensity of the spectra coming from the partitioned brine samples.

Figure 8.1 shows the signal intensity and peak integration values of 0.2 mM PABA dissolved in a $CaCl_2$ aqueous phase of 0.1%, 1%, 5%, 10%, and 100% salinities. It can be seen that the intensity of the NMR signals decreases as the conductivity of the aqueous solution increases, at the same concentration of PABA. The best fit through the experimental data consisted of an exponential function. As a result, the dependent variable of the function (Y) represents the intensity attenuation and its reciprocal gives the correction factor by which we have to multiply the signals generated at a given value of conductivity to obtain the corrected intensity. This correction

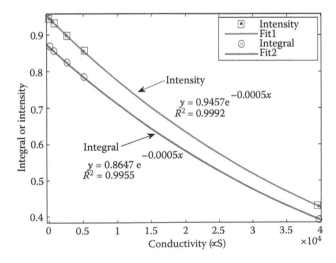

FIGURE 8.1 Intensity/peak integrals for the PABA NMR signals vs. aqueous-phase conductivity.

factor was subsequently used for all the aqueous solutions of NaCl and CaCl$_2$ with salinity values of 0.1%, 1%, 10%, and 100% and applied to the NMR spectra on the partitioned samples.

The region of the NMR spectra corresponding to NAs was determined by carrying out partitioning tests with a 1 wt.% solution of the commercial NA blend in decane and CaCl$_2$ solution at 1% and 100% salinity. Figure 8.2 shows that the NMR spectra for these acids fall in the 0–2 ppm region.

Figure 8.3 shows the NMR spectra of the partitioned brine with deasphalted GF (using the standard ASTM D2007-80) in comparison with the GF crude oil containing asphaltene, after 1 and 5 days. GF oil exhibits a viscosity of 43.5 cP and a density of 0.91 g/mL at 22°C with a C7-asphaltene content of 9 wt.%. As can be seen, the concentration of the naphthenic materials is much higher in the aqueous phase that is partitioned with the deasphalted GF especially after 5 days. These results indicate that adsorption of asphaltene in competition with naphthenic components can coat the interface to limit the amount of partitioned NA. This observation supports the hypothesis regarding different trends in the rate of NA/CaNA (calcium naphthenate) as concentration increases.

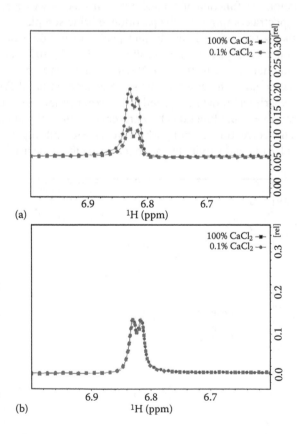

FIGURE 8.2 NMR spectra of the NA/decane partitioned with (a) 1% and (b) 100% salinity CaCl$_2$, respectively.

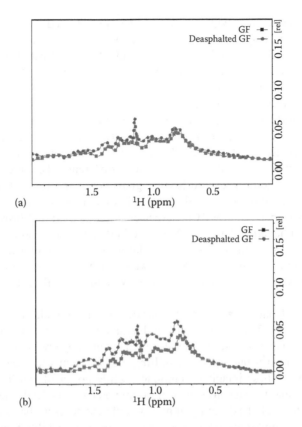

FIGURE 8.3 Partitioning test between deasphalted GF and (a) 1% CaCl₂ 1 day and (b) 1% CaCl₂ 5 days.

As described in the preceding paragraphs, ¹H-NMR spectroscopy was successfully used to determine the relative concentrations of naphthenic components partitioned from crude oil into the aqueous phase of different ionic strengths. At an early time, a higher concentration of organic components is detected in the brine with higher ionic strength.

8.5 OIL IN POROUS MEDIA

The technological evolution of the NMR technique in the last 10 years has allowed the study of oil in situ. NMR relaxation measurements and imaging have been used to study the fluid distribution within preserved core samples [51]. This type of study can be useful for the monitoring of imbibition processes in preserved cores at reservoir conditions during a waterflood, which is of particular interest in making production decisions. In fractured reservoirs, the timescale and magnitude of imbibition from the fractures into the matrix governs the viability of the reservoir.

Relaxation time (T_1) analysis and NMR imaging have also been used to determine the effect of aging in crude oil on the fluid distribution inside heterogeneous carbonate cores [52]. These experiments indicated that before aging, there is preferential oil distribution on large vugs, with only a slight amount situated in microporous areas. After aging, water evaporation allows an enhanced surface contact of the oil with the pore walls. NMR has also been applied to quantify the effect of sorbed crude oil on the spin-lattice relaxation time of water-saturated silica gels [53]. The results of this investigation indicate that ^{1}H-NMR is a potential means of detecting sorbed or residual hydrocarbons in natural environments.

The study of diffusion processes through NMR has become very popular, and many research efforts have been devoted to develop pulse sequences and special NMR probes to this end. The molecular diffusion coefficients of brine, oil, and gas molecules typically have values that are very different. Therefore, the diffusion attenuation of a suite of measured NMR signals contains sufficient information to allow the differentiation of these elements. This fact prompted the development of a magnetic resonance fluid (MRF) characterization method that involves the modeling of NMR relaxation data. The MRF method was used to study live and dead hydrocarbon mixtures, and it was shown to faithfully provide information regarding total porosity, bulk volume of irreducible water, brine and hydrocarbon saturation, hydrocarbon-corrected permeability, and oil viscosity in oil- and brine-saturated rocks [54,55].

The research frontier for oil and gas production is the understanding of unconventional reservoirs. The present focus in this area is the direct resolution of small-scale details in the rock system. Three-dimensional image reconstruction of shale has been performed using argon-ion milling and field emission scanning electron microscopy [56]. These imaging techniques, while powerful, can be significantly affected by the quality and consistency of sample preparation. Understanding molecular transport through the diversity of micro- and nanopores in shales requires other techniques [56]. Fluid transport in low- or ultra-low-permeability reservoirs can be non-Darcian, e.g., low-velocity non-Darcy flows [57]. On the other hand, Darcy's law, which is used to define permeability, assumes viscous flow through porous media. This reflects a bias arising from characterization of conventional rock, despite the fact that transport processes associated with gas flow in tight sands and shale are more complex than those associated with conventional reservoirs [58]. PFG NMR has been used to show that complex processes in nanoscale porous systems can be affected by phenomena such as evaporation–condensation [59]. This illustrates that a better understanding of the relationship between gas transport and pore space characteristics in shale and other low-permeability systems, including hydrophobicity or sorption properties of inorganic and organic support, is needed. Traditional gas permeability measurements have been shown to be insufficient to this end [60].

Sigal and Odusina [61] have shown through low-field NMR that the methane spin-lattice relaxation time T_2 value is consistent with sorption on hydrophobic organic pores, which can produce dipole–dipole interactions to relax magnetization in such a short time scale as opposed to values expected in the bulk phase. This surface relaxation can serve as an indicator of the fraction of methane stored in organic pores. On the other hand, water exhibiting a short T_2 value of the order of 1 ms has been inferred to arise from water residing in water-wet clays and inorganic surfaces, e.g., quartz. Despite all the advances in characterization technologies [62], the focus until now has been in evaluating traditional petrophysical indicators such as permeability and wettability, geochemical characterization, e.g., kerogen type, etc., to attempt to establish a connection with the potential of the shale gas reservoir. However, the impact of water injection through drilling and stimulation operations is still lacking. For instance, Ruppert et al. [63] have shown through the use of ultra-small-angle neutron scattering (ultra-SANS) and the more traditional SANS that most pores in a wide size range in shale samples studied in their work are accessible to both methane and water (in their deuterated forms). However, pores below 30 nm appear to be more accessible to water than methane. This observation indicates that water could decrease the pore space connectivity and therefore competes with methane transport, hence potentially producing a detrimental impact on gas production. Methane adsorption experiments have been conducted in clay-rich rocks as models of unconventional gas reservoirs [64]. The intent of the aforementioned study was to determine the distribution of gas between free gas and that adsorbed onto organic matrix or surfaces (kerogen) and clays. It has been argued that free gas is an important component for determining gas-in-place in shale reservoirs; however, adsorbed gas determines the longevity of shale gas production to some extent. Sorption isotherms at different temperatures can provide the sorption characteristics in shale gas rock [64]. Ji et al.'s work directly probed the sorption capacity onto different minerals in sedimentary rock. They showed that methane may preferentially adsorb on kerogen in contrast with clays. However, this important research work did not include the effect of moisture, although the authors explicitly mentioned the importance of water on sorption capacity.

8.6　OTHER APPLICATIONS

The introduction of new data analysis methods into the NMR realm has favored the study of crude oils. Silva et al. [65] indicated that NMR and its portfolio of techniques provide one of the most powerful analytical techniques for analysis of heavy crude oil mixtures. As aforementioned, improvements have been possible with the use of techniques such as distortionless enhancement by polarization transfer, GASPE, correlated spectroscopy, etc. As the authors described, complex information provided by NMR spectroscopy requires a statistical approach to fully use spectral data for characterization of crude oils. Moreover, the great challenge is to use NMR online, as a precise real-time analysis. The latter emphasizes the need for statistical techniques applied to complex spectral datasets. These techniques are encompassed in the so-called multivariate data analysis. Silva et al. highlight principal component analysis (PCA), PLS, and artificial neural networks (ANN). The main objective of PCA is to reduce the dimensionality of the dataset to aid in establishing the structure relationship between the independent variables (spectral data) and the predictions/observations (dependent variables), e.g., crude oil properties. PLS regression provides the correlation between the matrix of inputs and observations. Finally, ANN regression is a mathematical model that contributes to modeling the functional relationship between inputs and outputs. PLS is emerging as a powerful technique to predict the physicochemical properties of crude oil mixtures and spectroscopic datasets obtained from a variety of NMR techniques, including the use of multiple nuclei.

de Peinder et al. [66] conducted an important analysis using PLS regression combining infrared and NMR data. Although their results indicated that redundancy of data did not produce significant improvements in predictability of properties, their study brings these hybrid techniques to the mainstream analytical world. Additionally, Flumignan et al. [67] showed a powerful use of PLS regression as applied to ^{13}C-NMR data on the prediction of gasoline physicochemical properties. While this particular application focused on downstream uses, we view potential in extrapolation of these techniques to predict the properties of crude oil fractions as well as more complex systems such as water–crude oil macroemulsions.

An intriguing application of multivariate regression methods was presented by Zheng et al. [68]. Their intent was the simultaneous determination of water and oil content in oil sludge through PLS regression ^{1}H-NMR relaxometry and chemometrics. This latter term refers essentially to the use of PLS regression to correlate spectral data to component properties. They combined PCA and PLS to arrive at an appropriate PLS regression model to predict oil and water content simultaneously through analysis of T_2 relaxation time data. This resulted in a robust and accurate prediction based on a relatively simple calibration of data.

On the high-field side, NMR has been used successfully for direct imaging of porous systems since the inception of PFG techniques [69]. These techniques have been used to develop displacement propagators (for instance, see References 70 and 71). These techniques rely on homogeneous high magnetic fields and produce several-micron-resolution images. Among displacement propagators, NMR can directly provide diffusion propagators. In very recent times, researchers have theoretically demonstrated the ability to produce NMR-based diffusion pore images [72,73]. Traditional diffusion measurements in NMR rely on the application of symmetric consecutive field gradients in pulse sequences, namely $G_2 = -G_1$ of equal duration. This technique does not allow one to resolve diffusion images because reversion of field gradient wipes out the needed phase information. Laun et al. [72,73] have demonstrated that applying consecutive asymmetric magnetic field gradients that satisfy the condition $G_2t_2 = -G_1t_1$ can potentially provide the highly sought images of diffusion at the nanoscale and was naturally coined diffusion pore imaging. This technique does not produce spatially resolved images, but instead provides an average of the pore shape functions and their corresponding distribution density in the sample. The potential of such a technique is that if proven feasible in unconventional resource rock, it would directly relate gas diffusion to pore space geometrical characteristics, possibly leading to effective transport-weighted pore indicator. Shemesh et al. [74,75] experimentally demonstrated this diffusion imaging technique through the use of double-PFG NMR in several heterogeneous systems, including ground quartz samples. This technique is yet to be demonstrated in rocks of commercial interest.

Polymer flooding is a chemical-enhanced oil recovery process that consists of using water-soluble polymers of high molecular weight such as partially hydrolyzed polyacrylamide (HPAm), as water flooding additives in order to control displacing fluid mobility in reservoirs, to improve sweep efficiencies, and to increase oil production. Although not applied directly to crude oil or any of its derivatives, NMR has been used to characterize HPAm in the presence of Cr(III) acetate [76,77].

8.7 SUMMARY

We have shown that NMR has widespread applications in the oil and gas industry. These applications span the value chain to cover upstream characterization needs all the way to more traditional product characterization downstream. Analytical methods and NMR in particular have moved from niche areas in refining and petrochemistry to applications in production problems and porous medium characterization. Multivariate statistical analysis has substantially improved our ability to extract physicochemical information from complex spectral information from NMR spectroscopy. PCA and PLS are being increasingly used to build predictive models that aid characterization of crude oil and its fractions. Promising combinations of time-domain and high-field NMR are under development; however, recent published results indicate that these techniques will become routine in analytical laboratories. Finally, powerful relaxometry and diffusometry provide opportunities to enhance our understanding of oil and gas production in conventional and unconventional reservoirs. While this research frontier in NMR is only taking its first step, results today signal game-changing opportunities for the applications of NMR.

REFERENCES

1. Alvarado, V., Wang, X., and Moradi, M. 2011. Stability proxies for water-in-oil emulsions and implications in aqueous-based enhanced oil recovery. *Energies* 4:1058–1086.
2. Moradi, M., Topchiy, E., Lehmann, T. E., and Alvarado, V. 2013. Impact of ionic strength on partitioning of naphthenic acids in water-crude oil systems—Determination through high-field NMR spectroscopy. *Fuel* 112:236–248.
3. Pedersen, W. B., Hansen, A. B., Larsen, E., and Neilsen, B. 1991. Wax precipitation from North Sea crude oils. 2. Solid-phase content as function of temperature determined by pulsed NMR. *Energy Fuels* 5:908–913.
4. Sato, S., Matsumura, A., Urushigawa, Y., Urushigawa, M., and Al-Muzaini, S. 1998. Structural analysis of weathered oil from Kuwait's environment. *Environ. Int.* 24: 77–87.
5. Smirnov, M. B., and Vanyukova, N. A. 2012. ^{13}C NMR identification in crude oils of hydrocarbons bearing ethyl and *n*-alkyl substituents on their main chain. *Petrol. Chem.* 52:383–391.
6. Goldstein, T. P., and Schmitt, K. D. 1994. Carbon centre types as determined by MAS–CP NMR on frozen fossil fuels. *Fuel* 73:397–403.
7. Ali, F., Khan, S. A., and Ghaloum, N. 2004. Structural studies of vacuum gas oil distillate fractions of Kuwaiti crude oil by nuclear magnetic resonance. *Energy Fuels* 18:1798–1805.
8. McKay, J. F., Amend, P. J., Harnsberger, P. M., Cogswell, T. E., and Latham, D. R. 1981. Composition of petroleum heavy ends. I. Separation of petroleum >675°C residues. *Fuel* 60:14–16.
9. McKay, J. F., Harnsberger, P. M., Erickson, R. B., Cogswell, T. E., and Latham, D. R. 1981. Composition of petroleum heavy ends. II. Characterization of compound types in petroleum >675°C residues. *Fuel* 60:17–26.
10. McKay, J. F., Latham, D. R., and Haines, W. E. 1981. Composition of petroleum heavy ends. III. Comparison of the composition of high-boiling petroleum distillates and petroleum >675°C residues. *Fuel* 60:27–32.
11. Hasan, M. U., Ali, M. F., and Bukhari, A. 1983. Structural characterization of Saudi Arabian heavy crude oil by NMR spectroscopy. *Fuel* 62:518–523.

12. Hasan, M. U., Bukhari, A., and Ali, M. F. 1985. Structural characterization of Saudi Arabian heavy crude oil by NMR spectroscopy. *Fuel* 64:839–841.

13. Sharma, B. K., Tyagi, O. S., Aloopwan, M. K. S., and Bhagat, S. D. 2000. Spectroscopic characterization of solvent soluble fractions of petroleum vacuum residues. *Pet. Sci. Technol.* 18:249–272.

14. Smirnov, M. B., Poludetkina, E. N., Vanyukova, N. A., and Parenago, O. P. 2010. Comparative ¹³C NMR analysis of the composition of saturated petroleum and bitumen-oid hydrocarbons: Potentialities and outlook. *Petrol. Chem.* 51:118–127.

15. Castro, A. T. 2006. NMR and FTIR characterization of petroleum residues: Structural parameters and correlations. *J. Braz. Chem. Soc.* 17:1181–1185.

16. Cookson, D. J., Lloyd, C. P., and Smith, B. E. 1986. A novel semi-empirical relationship between aromaticities measured from ¹H and ¹³C N.M.R. spectra. *Fuel* 65:1247–1253.

17. Martos, C., Coto, B., Espada, J. J., Robustillos, M. D., Gómez, S., and Peña, J. L. 2008. Experimental determination and characterization of wax fractions precipitated as a function of temperature. *Energy Fuels* 22:708–714.

18. Li, S.-P., Tang, F., Shen, B.-X., Zhang, B.-L., Yang, J.-Y., and Xu, X.-R. 2008. A study of the characteristics in blended crudes at the optimal activation state: Part I. *Pet. Sci. Technol.* 26:1499–1509.

19. Nielsen, K. E., Dittmer, J., Malmendal, A., and Nielsen, N. C. 2008. Quantitative analysis of constituents in heavy fuel oil by ¹H nuclear magnetic resonance (NMR) spectroscopy and multivariate data analysis. *Energy Fuels* 22:4070–4076.

20. Masili, A., Puligheddu, S., Sassu, L., Scano, P., and Lai, A. 2012. Prediction of physical–chemical properties of crude oils by ¹H NMR analysis of neat samples and chemometrics. *Magn. Reson. Chem.* 50:729–738.

21. Filippelli, L., Gentile, L., Rossi, C. O., Ranieri, G. A., and Antunes, F. 2012. Structural change of bitumen in the recycling process by using rheology and NMR. *Ind. Eng. Chem. Res.* 51:16346–16353.

22. Woods, J., Kung, J., Kingston, D., Kotlyar, L., Sparks, B., and McCracken, T. 2008. Canadian crudes: A comparative study of SARA fractions from a modified HPLC separation technique. *Oil Gas Sci. Technol.* 63:151–163.

23. Pereira, J. F. B., Gudiña, E. J., Costa, R., Vitorino, R., Teixeira, J. A., Coutinho, J. A. P., and Rodrigues, L. 2013. Optimization and characterization of biosurfactant production by *Bacillus subtilis* isolates towards microbial enhanced oil recovery applications. *Fuel* 111:259–268.

24. Khadim, M. A., and Sarbar, M. A. 1999. Role of asphaltene and resin in oil field emulsions. *J. Pet. Sci. Eng.* 23:213–221.

25. Buenrostro-Gonzalez, E., Espinosa-Peña, M., Andersen, S. I., and Lira-Galeana, C. 2001. Characterization of asphaltenes and resins from problematic Mexican crude oils. *Pet. Sci. Technol.* 19:299–316.

26. Luo, P., Wang, X., and Gu, Y. 2010. Characterization of asphaltenes precipitated with three light alkanes under different experimental conditions. *Fluid Phase Equilibr.* 291: 103–110.

27. Wang, X., and Gu, Y. 2011. Characterization of precipitated asphaltenes and deasphalted oils of the medium crude oil–CO₂ and medium crude oil–*n*-pentane systems. *Energy Fuels* 25:5232–5241.

28. Gaspar, A., Zellermann, E., Lababidi, S., Reece, J., and Schrader, W. 2012. Impact of different ionization methods on the molecular assignments of asphaltenes by FT-ICR mass spectrometry. *Anal. Chem.* 84:5257–5267.

29. Calemma, V., Iwanski, P., Nali, M., Scotti, R., and Montanari, L. 1995. Structural characterization of asphaltenes of different origins. *Energy Fuels* 9:225–230.

30. Siddiqui, M. N., and Ali, M. F. 1999. Investigation of chemical transformations by NMR and GPC during the laboratory aging of Arabian asphalt. *Fuel* 78:1407–1416.

31. Pekerar, S., Lehmann, T. E., Méndez, B., and Acevedo, S. 1999. Mobility of asphaltene samples studied by ^{13}C NMR spectroscopy. *Energy Fuels* 13:305–308.
32. Kapur, G. S., Findeisen, M., and Berger, S. 2000. Analysis of hydrocarbon mixtures by diffusion-ordered NMR spectroscopy. *Fuel* 79:1347–1351.
33. Durand, E., Clemancey, M., Quoineaud, A.-A., Verstraete, J., Espinat, D., and Lancelin, J.-M. 2008. [1]H diffusion-ordered spectroscopy (DOSY) nuclear magnetic resonance (NMR) as a powerful tool for the analysis of hydrocarbon mixtures and asphaltenes. *Energy Fuels* 22:2604–2610.
34. Durand, E., Clemancey, M., Lancelin, J.-M., Verstraete, J., Espinat, D., and Quoineaud, A.-A. 2009. Aggregation states of asphaltenes: Evidence of two chemical behaviors by ^{1}H diffusion-ordered spectroscopy nuclear magnetic resonance. *J. Phys. Chem.* 113: 16266–16276.
35. Hauser, A., Bahzad, D., Stanislaus, A., and Behbahani, M. 2008. Thermogravimetric analysis studies on the thermal stability of asphaltenes: Pyrolysis behavior of heavy oil asphaltenes. *Energy Fuels* 22:449–454.
36. Hirasaki, G. J., Miller, C. A., Raney, O. G., Poindexter, M. K., Nguyen, D. T., and Hera, J. 2011. Separation of produced emulsions from surfactant enhanced oil recovery processes. *Energy Fuels* 25:555–561.
37. Mandal, A., Samanta, A., Bera, A., and Ojha, K. 2010. Characterization of oil–water emulsion and its use in enhanced oil recovery. *Ind. Eng. Chem. Res.* 49:12756–12761.
38. Abraham, R. J., Fisher, J., and Loftus, P. 1988. *Introduction to NMR Spectroscopy*. Hoboken: John Wiley & Sons.
39. Sjöblom, J., Aske, N., Auflem, I. H., Brandal, Ø., Havre, T. E., Saether, O., Westvik, A., Johnsen, E. E., and Kallevik, H. 2003. Our current understanding of water-in-crude oil emulsions. Recent characterization techniques and high pressure performance. *Adv. Colloid Interface Sci.* 100:399–473.
40. Tanner, J. E., and Stejskal, E. O. 1968. Restricted diffusion of protons in colloidal systems by the pulsed field gradient method. *J. Phys. Chem.* 49:1768–1777.
41. Johns, M. L. 2009. NMR studies of emulsions. *Curr. Opin. Colloid Interface Sci.* 14:178–183.
42. Murday, J. S., and Cotts, J. M. 1968. Self-diffusion coefficient of liquid lithium. *J. Chem. Phys.* 48:4938–4945.
43. Packer, K. J., and Rees, C. 1972. Pulsed NMR studies of restricted diffusion. I. Droplet size distributions in emulsions. *J. Colloid Interface Sci.* 40:206–218.
44. Peña, A. A., and Hirasaki, G. J. 2003. Enhanced characterization of oilfield emulsions via NMR diffusion and transverse relaxation experiments. *Adv. Colloid. Interface Sci.* 105:103–150.
45. Garasanin, T., Cosgrove, T., Marteaux, L., Kretschmer, A., Goodwin, A., and Zick, K. 2002. NMR self-diffusion studies on PDMS oil-in-water emulsion. *Langmuir* 18:10298–10304.
46. Honorato, H. D. A., Silva, R. C., Piumbini, C. K., Zucolotto, C. G., de Souza, A. A., Cunha, A. G., Emmerich, F. G., Lacerda Jr., V., de Castro, E. V. R., Bonagamba, T. J., and Freitas, J. C. C. 2012. ^{1}H low- and high-field NMR study of the effects of plasma treatment on the oil and water fractions in crude heavy oil. *Fuel* 92:62–68.
47. Hemmingsen, P. V., Silset, A., Hannisdal, A., and Sjöblom, J. 2005. Emulsions of heavy crude oils. I: Influence of viscosity, temperature, and dilution. *J. Disper. Sci. Technol.* 26:615–627.
48. Less, S., Hannisdal, A., and Sjöblom, J. 2008. An electrorheological study on the behavior of water-in-crude oil emulsions under influence of a DC electric field and different flow conditions. *J. Disper. Sci. Technol.* 29:106–114.
49. Opedal, N., Sørland, G., and Sjöblom, J. 2010. Emulsion stability studied by nuclear magnetic resonance (NMR). *Energy Fuels* 24:3628–3633.

50. Knudsen, A., Nordgård, E. L., Diou, O., and Sjöblom, J. 2012. Methods to study naphthenate formation in w/o emulsions by the use of a tetraacid model compound. *J. Disper. Sci. Technol.* 33:1514–1524.
51. Davies, S., Hardwick, A., Roberts, D., Spowage, K., and Packer, K. J. 1994. Quantification of oil and water in preserved reservoir rock by NMR spectroscopy and imaging. *Magn. Reson. Imaging* 12:349–353.
52. Maddinelli, G., and Vitali, R. 1998. Evaluation of chemically-induced pore surface modifications on rock cores. *Magn. Reson. Imaging* 16:669–672.
53. Daughney, C. J., Bryar, T. R., and Knight, R. J. 2000. Detecting sorbed hydrocarbons in a porous medium using proton nuclear magnetic resonance. *Environ. Sci. Technol.* 34:332–337.
54. Freedman, R., Lo, S., Flaum, M., Hirasaki, G. J., Matteson, A., and Sezginer, A. 2001. A new NMR method of fluid characterization in reservoir rocks: Experimental confirmation and simulation results. *Soc. Petrol. Eng. J.* 6:452–464.
55. Freedman, R., Heaton, N., and Flaum, M. 2002. Field application of a new nuclear magnetic resonance fluid characterization method. *Soc. Petrol. Eng. J.* 5:455–463.
56. Slatt, R. M., and O'Brien, N. R. 2011. Pore types in the Barnett and Woodford gas shales: Contribution to understanding gas storage and migration pathways in fine-grained rocks. *AAPG Bull.* 95:2017.
57. Zhu, W., Song, H., Huang, X., Liu, X., He, D., and Ran, Q. 2011. Pressure characteristics and effective deployment in a water-bearing tight gas reservoir with low-velocity non-Darcy flow. *Energy Fuels* 25:1111.
58. Fathi, E., and Akkutlu, I. Y. C. 2009. Matrix heterogeneity effects on gas transport and adsorption in coalbed and shale gas reservoirs. *Transport Porous Med.* 80:281.
59. Carrara, C., Pagès, G., Delaurent, C., Veil, S., and Caldarelli, S. 2011. Mass transport of volatile molecules in porous materials: Evaporation–condensation phenomena described by NMR diffusometry. *J. Phys. Chem.* 115:18776–18781.
60. Cui, X., Bustin, A. M. M., and Bustin, R. M. 2009. Measurements of gas permeability and diffusivity of tight reservoir rocks: Different approaches and their applications. *Geofluids* 9:208.
61. Sigal, R. F., and Odusina, E. 2011. Laboratory NMR measurements of methane saturated Barnett shale samples. *Petrophysics* 52:32–49.
62. Bai, B., Elgmati, M., Zhang, H., and Wei, M. 2012. Rock characterization of Fayetteville shale gas plays. *Fuel* 105:645:652.
63. Ruppert, L. F., Sakurovs, R., Blach, T. P., He, L., Melinichenko, Y. B., Mildner, D. F. R., and Alcantar-Lopez, L. 2013. A USANS/SANS study of the accessibility of pores in the Barnett shale to methane and water. *Energy Fuels* 27:772–779.
64. Ji, L., Zhang, T., Milliken, K. L., Qu, J., and Zhang, X. 2012. Experimental investigation of main controls to methane adsorption in clay-rich rocks. *Appl. Geochem.* 27:2533–2545.
65. Silva, S. L., Silva, A. M. S., Ribeiro, J. C., Martins, F. G., Da Silva, F. A., and Silva, C. M. 2011. Chromatographic and spectroscopic analysis of heavy crude oil mixtures with emphasis in nuclear magnetic resonance spectroscopy: A review. *Anal. Chim. Acta* 7070:18–37.
66. de Peinder, P., Visser, T., Petrauskas, D. D., Salvatori, F., Soulimani, F., and Weckhuysen, B. M. 2009. Partial least squares modeling of combined infrared, ^{1}H NMR and ^{13}C NMR spectra to predict long residue properties of crude oils. *Vib. Spectrosc.* 51:205–212.
67. Flumignan, D. L., Sequinel, R., Hatanaka, R. R., Boralle, N., and de Olivera, J. E. 2012. Carbon nuclear magnetic resonance spectroscopic profiles coupled to partial least-squares multivariate regression for prediction of several physicochemical parameters of Brazilian commercial gasoline. *Energy Fuels* 26:5711–5718.

68. Zheng, X., Jin, Y., Chi, Y., and Ni, M. 2013. Simultaneous determination of water and oil in oil sludge by low-field ^{1}H NMR relaxometry and chemometrics. *Energy Fuels* 27:5787–5792.
69. Hardy, E. H. 2012. *NMR Methods for the Investigation of Structure and Transport.* Berlin: Springer-Verlag.
70. Hussain, R., Pintelon, T. R. R., Mitchell, J., and Johns, M. L. 2010. Using NMR displacement measurements to probe CO_2 entrapment in porous media. *AIChE J.* 57:1700.
71. Kutsovsky, Y., Alvarado, V., Scriven, L. E., Davis, H. T., and Hammer, B. 1996. Dispersion of paramagnetic tracers in bead packs by T1 mapping: Experiments and simulations. *Magn. Reson. Imaging* 14:833.
72. Laun, F. B., Kuder, T. A., Semmler, W., and Stieltjes, B. 2011. Determination of the defining boundary in nuclear magnetic resonance diffusion experiments. *Phys. Rev. Lett.* 107:048102.
73. Laun, F. B., Kuder, T. A., Wetscherek, A., and Stieltjes, B. 2011. NMR diffusion pore imaging. *Phys. Rev. E* 86:021906.
74. Shemesh, N., Adiri, T., Cohen, Y. 2011. Probing microscopic architecture of opaque heterogeneous systems using double-pulsed-field-gradient NMR. *J. Am. Chem. Soc.* 133:6028.
75. Shemesh, N., Ozarslan, E., Adiri, T., Basser, P. J., and Cohen, Y. 2010. Noninvasive bipolar double-pulsed-field-gradient NMR reveals signatures for pore size and shape in polydisperse, randomly oriented, inhomogeneous porous media. *J. Phys. Chem.* 133:044705.
76. Vargas-Vasquez, S. M., Romero-Zerón, L. B., and MacMillan, B. 2007. ^{1}H NMR characterization of HPAm/Cr(III) acetate polymer gel components. *Int. J. Polym. Anal. Ch.* 12:115–129.
77. Vargas-Vasquez, S. M., Romero-Zerón, L. B., Macgregor, R., and Gopalakrishnan, S. 2007. Monitoring the cross-linking of a HPAm/Cr(III) acetate polymer gel using ^{1}H NMR, UV spectrophotometry, bottle testing, and rheology. *Int. J. Polym. Anal. Ch.* 12:339–357.

Section V

Heavy Ends and Asphaltenes

9 On-Column Filtration Asphaltene Characterization Methods for the Analysis of Produced Crude Oils and Deposits from Upstream Operations

Estrella Rogel, César Ovalles, and Michael E. Moir

CONTENTS

CONTEXT

- Asphaltenes are considered "bad actors" when crude oils and their fractions are produced, transported, or refined
- Loss production by asphaltene plugging could yield significant financial effects for upstream operations

ABSTRACT

In the last few years, we have developed three new analytical techniques for asphaltene characterization: determination of asphaltene content by on-column filtration, asphaltene solubility profile, and separation of asphaltenes in solubility fractions. The results showed that the on-column filtration can be effectively and rapidly used for determining asphaltene content of crude oil samples. Similarly, the asphaltene solubility profile method can be effectively utilized to measure the stability of extra heavy oil, crude oil/naphtha blends, and crude oil during CO_2 flooding, to screen additives to improve asphaltene solubility, and to determine the enrichment of low solubility asphaltenes found in solid deposits. Similarly, the asphaltene solubility fraction method is useful to monitor changes in asphaltene distribution during steam injection processes. These examples show the potential of these techniques to provide useful information about changes in asphaltene behavior during production operations and for the selection and design of treatments and recovery processes.

9.1   INTRODUCTION

Asphaltenes are defined as the fraction of petroleum that is insoluble in n-paraffins (e.g., C5, C7, etc.) but soluble in aromatics (toluene, xylene, etc.). They are considered "bad actors" when crude oils and their fractions are produced, transported, or refined. They have been "blamed" for several operational issues throughout the petroleum value chain such as [1,2]

- Reducing the permeability of oil-bearing formations and the concomitant reduction in the volumetric production of wells and percentages of oil recovery
- Increased cost for lifting and transportation due to the high viscosity of asphaltenic crude oils
- Plugging of tubing, pumps, valves, and pipelines due to asphaltene precipitation
- Causing fouling in heat exchangers and other devices in production and refinery operations
- Reducing and limiting the yield of residue conversion during catalytic upgrading processes
- Contributing to catalyst poisoning through coke and metal deposition
- Being the source of coke during thermal upgrading processes

Specifically in upstream operations, the loss in production due to asphaltene plugging could yield financial losses on the order of tens of millions of US dollars, depending on the field that it is being produced. For example, in the Prinos petroleum production field in Greece, some wells completely ceased flowing in a matter of a few days after an initial production rate of up to 3000 BPD [3]. Other examples of solid precipitation were reported by Carbognani et al. [4] during oil production in Venezuelan fields. The authors identified remedial actions based on deposit characterization and developed criteria for selection of the most efficient dispersant additives and solvent mixtures for cleaning production strings [4].

Asphaltene deposition problems during production operations originate from a decrease in the solubility of asphaltenes. As asphaltenes become less soluble, they form floccules that, under the right conditions, can deposit blocking pathways. In production operations, the decrease in the solubility of asphaltenes is normally related to the decreasing solvent power (ability to dissolve asphaltenes) of the maltenes. Changes in pressure and temperature during the natural declining of production can cause asphaltene precipitation. However, it can also be caused by recovery process and treatments. For example, during carbon dioxide floods and when this gas is mixed with the crudes, asphaltenes can become destabilized and may deposit on the mineral formation or elsewhere in the production system [5]. There are several field examples in which no asphaltene deposition was experienced during primary production. However, once CO_2 was injected downhole, severe precipitation issues were observed [6–8]. For example, for the CO_2 pilot in an onshore Abu Dhabi Field, three asphaltene removal jobs were carried out during the life of the pilot (1.5 years). Asphaltene build-up was observed after only 3–4 weeks after cleanup [5]. It is also well known that acidification treatments and other stimulation procedures can modify the solvent power of the maltenes, and therefore induce precipitation. For example, it has been reported that iron-contaminated acid promoted the precipitation of asphaltenes when acidizing certain oil-bearing zones [9].

There are several techniques that can be used to deal with this issue: (i) control of pressure or other operational variables such as comingling, (ii) periodic cleaning (either chemical or mechanical), and (iii) additive injection to prevent asphaltene precipitation.

In 2001, it was reported that each asphaltene cleanup operation using organic solvents with a coiled tubing unit costs, on average, US$200,000 onshore [4]. More recently, in 2012, it has been estimated that each remediation event costs around $500,000 (onshore) or $3,000,000 (offshore) [10]. In fact, in some cases, owing to the severity of asphaltene deposition and based on the economics of the cleanup operation, it was recommended to implement continuous downhole chemical injection of asphaltene inhibitors in producer wells [6].

Additionally, it has been reported that the presence of asphaltenes and its tendency to form aggregates plays a fundamental role in the high viscosity (~100,000 cP at room temperature) observed for heavy and extra heavy crude oils (H/XH crudes) [11]. Specifically, it has been shown that by removing the asphaltenes from these materials, a reduction of 2 to 3 orders of magnitude on the viscosity of the maltenes is obtained over the temperature range from 40°C to 80°C [9]. Also, asphaltenes interact extensively within themselves and with other similar molecules forming extended networks, and thus the viscosity of asphaltene-containing petroleum samples shows an exponential increase with their concentration [12]. Several routes have been investigated in the literature to overcome this issue, such as the use of emulsions [13–15], diluents [11,16], heated pipelines [17], carbon dioxide [18,19], precipitation and separation of asphaltenes [20,21], and use of additives [22–24]. In these routes, caution should be taken in maintaining the asphaltenes in solution so no solid precipitation problems are developed during the transportation and storage of H/XH crudes [11].

It is clear from the previous paragraphs that it is highly desirable for the petroleum industry to develop methods that can be used to accurately evaluate the effect that different recovery processes and chemical treatments (i.e., acidification, diluents, or additives) might have on asphaltene behavior.

One important step toward this end, namely understanding, mitigating, and solving asphaltene precipitation and transportation issues, is the analysis and characterization of the asphaltene fractions present in crude oils and solid deposits. With this goal in mind, in the last few years, we have developed three techniques for asphaltene analysis based on on-column filtration methods [25–29]. In this work, we demonstrate the usefulness of these methods in characterizing produced crude oils and deposits from upstream operations. These analytical techniques are as follows: determination of asphaltene content by the on-column filtration method [25,26], asphaltene solubility profile [27,28], and separation of asphaltenes in solubility fractions [29].

9.2 EXPERIMENTAL

9.2.1 MATERIALS

The high-performance liquid chromatography–grade solvents methylene chloride, methanol, n-heptane, and tetrahydrofuran were purchased from Fisher Scientific and used without further purification. A wide range of petroleum samples was tested with different API (American Petroleum Institute) gravities (from 8°API to 28°API) and from different parts of the world. Included were crude oils from Venezuela, Canada, California, Alaska, Mexico, Ecuador, and the Middle East. The extra heavy crude oil came from the South America (8.8°API), and the naphtha used for blending came from Chevron's Richmond Refinery [11]. Solid deposits and parent crude oils came from several Chevron upstream operations.

Gravimetric asphaltene determinations were carried out using the modified ASTM (American Society for Testing and Materials) D-6560 [30]. The three methods for asphaltene characterization have been described in detail in previous publications and will be briefly described in Sections 9.2.2–9.2.4 [25–29].

9.2.2 ASPHALTENE CONTENT BY THE ON-COLUMN FILTRATION METHOD

The detailed procedure for the determination of asphaltene content using the on-column filtration method has been described elsewhere [25,26]. In a typical analysis,

0.1 g of a petroleum sample is dissolved in methylene chloride and injected into a column packed with an inert material using n-heptane as the mobile phase. The first peak corresponds to the eluted maltenes (heptane solubles at 2–3 min). At 10 min, the solvent is switched from pure n-heptane to 90:10 methylene chloride/methanol and the second peak corresponds to the eluted asphaltenes at ~12 min. As reported, a calibration curve was prepared using asphaltenes extracted from virgin samples and asphaltenes were quantified using an evaporative light scattering detector (ELSD) [25,26].

9.2.3 SOLUBILITY PROFILE OF ASPHALTENES BY ON-COLUMN METHOD

A solution of the sample in methylene chloride is injected in a column packed with an inert material using n-heptane as the mobile phase. The first eluted fraction from the column is the maltenes, which are soluble in n-heptane. After this fraction has fully eluted, the mobile phase is changed gradually from pure n-heptane to 90:10 methylene chloride/methanol and then to 100% methanol. This procedure gradually redissolves the asphaltenes from the easy to dissolve (low solubility parameter) to the difficult to dissolve (high solubility parameter). As before, asphaltenes were quantified using an ELSD [27,28].

In general, asphaltenes from unstable materials are characterized by wider solubility profile distributions than asphaltenes from stable materials. A parameter for the distributions (ΔPS) was developed to quantify the asphaltene distributions [27,28]. The ΔPS was calculated from the cumulative areas of the whole distribution that represents the solubility profile of the asphaltenes in the sample. The cumulative areas were calculated using Excel by Microsoft Co., by applying the trapezoidal rule. Then, the ΔPS was determined by subtracting the time that corresponds to the 25% of area from the time that corresponds to the 75% of area [27,28].

9.2.4 SEPARATION OF ASPHALTENES IN SOLUBILITY FRACTIONS

As in the solubility profile analysis, the sample is dissolved in methylene chloride and injected in a column packed with an inert material using n-heptane as the mobile phase. Maltenes (heptane solubles) elute from the column as the first peak. The asphaltenes remained precipitated in the column and were fractionated according to their respective solubilities by switching the mobile phase in successive steps to solvents of increasing solubility parameters: (i) 10 min after the injection of the sample, the mobile phase was switched to a blend of 15% methylene chloride/85% n-heptane (solubility parameter of 16.05 MPa$^{0.5}$); (ii) 20 min after the injection of the sample, the mobile phase was switched to a blend of 30% methylene chloride/70% n-heptane (solubility parameter of 18.8 MPa$^{0.5}$); (iii) 30 min after the injection of the sample, the mobile phase was switched to 100% methylene chloride (solubility parameter of 20.3 MPa$^{0.5}$); and (iv) 40 min after the injection of the sample, the mobile phase was switched to a blend of 10% methanol/90% methylene chloride (solubility parameter of 21.23 MPa$^{0.5}$). After 10 additional min, the solvent was switched again to n-heptane. The eluted fractions were quantified using an ELSD as described elsewhere [29].

9.3 RESULTS AND DISCUSSION

As mentioned, we present in this work the use of three new characterization techniques: determination of asphaltene content by the on-column filtration method [25,26], asphaltene solubility profile [27,28], and separation of asphaltenes in solubility fractions [29], for the characterization of crude oils and deposits from upstream operations.

9.3.1 ASPHALTENE CONTENT OF PRODUCED CRUDE OILS

In the first example, asphaltene content was determined by using the on-column filtration method in nine virgin crude oils produced from different parts of the world (see Section 9.2). This method involves using mixtures of *n*-heptane, methylene chloride, and methanol [25,26]. After injecting the sample, the first peak corresponds to the maltenes (*n*-heptane solubles) and, after switching to 10% methanol in methylene chloride, the asphaltene peak (*n*-heptane insolubles) is determined using an ELSD. A typical example of the liquid chromatography (LC) trace can be found in Figure 9.1.

This on-column filtration technique requires a small amount of sample and takes only 20 min to perform. Additionally, it was found to have better reliability, repeatability, and reproducibility ($\pm 7\%$) than the traditional gravimetric methods ($\pm 14.5\%$). This makes it a good candidate for the monitoring of upstream processes [25,26].

In effect, the comparison between the determination of asphaltenes by the on-column method and the conventional gravimetric method, ASTM D-6560 [30], can be seen in Figure 9.2. In this figure, nine crude oils produced from different parts of the world are analyzed. As shown, there is a reasonable good correlation between both methods with an R^2 of ~0.90.

As previously reported [25,26], values obtained using the on-column method are larger since the test is run at room temperature conditions, while the modified ASTM

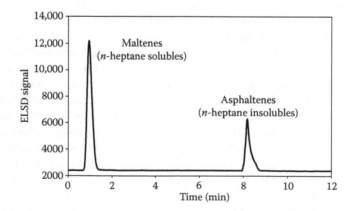

FIGURE 9.1 Typical LC trace of the on-column filtration determination of asphaltene concentration.

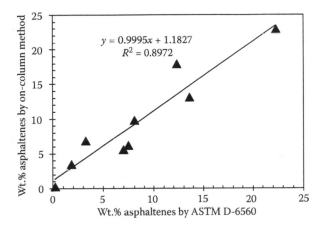

FIGURE 9.2 Comparison of the determination of asphaltenes by the on-column filtration method [10] and the conventional gravimetric method ASTM D-6560.

D-6560 uses a relatively higher temperature of around 80°C [30]. These results indicated that the determination of asphaltene content by using the on-column filtration method can be effectively done for analyzing crude oil samples from upstream operations. This method has been found very useful for analyzing samples from core extractions and from displacement tests in which sample volumes are not enough to carry out gravimetric asphaltene determinations.

9.3.2 Evaluation of Diluents and Additives Using Solubility Profile Measurements

The second method applied to the characterization of produced crude oils is the asphaltene solubility profile. This method allows determining the asphaltene precipitation tendency of crude oils and related products based on the evaluation of the solubility profile or solubility distribution of asphaltenes [27,28]. As in the previous method, the first eluted fraction from the column is the maltenes, which are soluble in n-heptane. For simplification, the maltene peak is not shown in Figure 9.3. After this fraction has eluted, the mobile phase is changed gradually from pure n-heptane to 90:10 methylene chloride/methanol and then to 100% methanol. A typical solubility profile of a produced crude oil is shown in Figure 9.3. This procedure gradually redissolves the asphaltenes from the easy to dissolve (low solubility parameter) to the difficult to dissolve (high solubility parameter) [27,28].

In general, asphaltenes from unstable materials are characterized by wider solubility profile distributions than asphaltenes from stable materials. A parameter for quantification of the distributions (ΔPS) was developed by subtracting the time that corresponds to the 25% of area from the time that corresponds to the 75% of area (see Figure 9.3) [24]. This parameter was correlated with the p-value determined by traditional flocculation onset measurements [27,28]. In general, larger values of

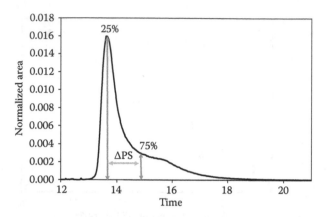

FIGURE 9.3   Typical LC trace of the asphaltene solubility profile of a virgin crude oil and how the ΔPS parameter is calculated.

ΔPS indicate a less stable material. A ΔPS >2.50 indicates stability problems, while values <2.00 indicate stable materials. In general, the repeatability of this method is ±0.05 [27,28].

This method for determining asphaltene solubility profiles requires a small amount of sample, takes only 35 min to perform, and can work for asphaltene concentrations as low as 500 ppm [24]. It has been used successfully for the analysis of a large variety of samples from virgin crude oils to highly processed petroleum fractions [10]. In this example, this method was used to determine the stability of extra heavy oil/diluent blends.

As mentioned in Section 9.1, the presence of asphaltenes and their tendency to form aggregates play a fundamental role in the high viscosity observed for H/XH crudes [11,12]. For these reasons, the use of diluents for transportation of H/XH crudes represents an attractive alternative [11–16]. However, caution should be taken in choosing the appropriate diluent so asphaltenes are not destabilized. To monitor this phenomenon, the asphaltene solubility profile method was used and the results are presented in Figure 9.4. As can be seen, the original extra heavy oil (8.8°API gravity) has a ΔPS of ~1.7, whereas extra heavy oil/naphtha blends (Figure 9.4) are stable with ΔPS <1.5. In fact, the addition of naphtha seems to improve the stability of the extra heavy oil at all concentrations tested (10%–90% vol.).

As discussed, there are several techniques that can be used to control and mitigate asphaltene precipitation and transportation issues. One of them is injection of additives downhole. In this example, the asphaltene solubility profile was used to screen and select additives to improve asphaltene solubility during surface production operations. Specifically, three different additives (X-1, X-2, and X-3) were used at concentrations between 15 and 600 ppm with respect to crude oil #3. Figure 9.5 shows the comparison between the original crude oil #3 (continuous trace) and the addition of 150 ppm of additive X-2 (discontinuous trace). As the figure shows, changes in the

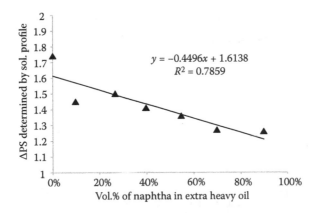

FIGURE 9.4 Stability parameter ΔPS for naphtha/extra heavy crude oil blends.

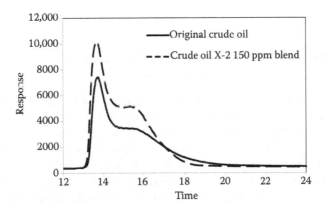

FIGURE 9.5 Asphaltene solubility profile of crude oil #3 in the absence and presence of 150 ppm additive X-2.

solubility profile of the asphaltenes are clearly observed. In effect, in the presence of X-1, the ΔPS of the sample is slightly reduced from 2.1 to 2.0 in the presence of the additive, indicating an increase in asphaltene stability [27,28].

Figure 9.6 shows the parameter ΔPS as a function of the concentration of the different additives, X-1, X-2, and X-3. As shown, there are increases in the solubility of the asphaltene as a consequence of the addition of the additives in comparison with the original crude oil. In this particular crude oil, additive X-1 seems to be the most effective (lowest ΔPS at 1.4) in increasing the solubility of the asphaltenes in the crude oil and thus its stability to solid precipitation [27,28]. Another interesting aspect, in terms of the effectiveness of the additives, is that after a certain concentration threshold, additive activity seems to remain almost constant, indicating that larger concentrations do not necessarily induce significantly better performance.

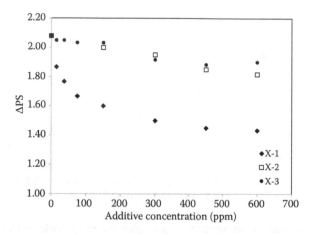

FIGURE 9.6 Stability parameter ΔPS for crude oil #3 in function of the concentration of additives X-1, X-2, and X-3.

9.3.3 EFFECT OF ENHANCED OIL RECOVERY PROCESSES ON ASPHALTENE BEHAVIOR AND COMPOSITION

As also mentioned in Section 9.1, carbon dioxide has been used for downhole injection in enhanced oil recovery processes (EORs) and as diluent for H/XH crudes transportation [11,16]. However, during CO_2 floods, asphaltenes can become destabilized and may deposit on the mineral formation or elsewhere in the production system [5]. As in the previous example, the asphaltene solubility profile method was used to monitor this phenomenon, and the results are presented in Figure 9.7. In the figure, two crude oils from the same oilfield are compared. Crude oil #1 (discontinuous trace) comes from a section of the oilfield produced by CO_2 flooding, whereas

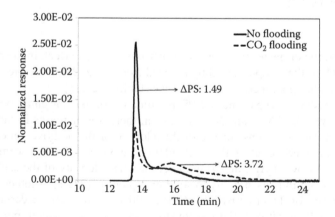

FIGURE 9.7 Stability parameter ΔPS for the original crude oil #2 and the same crude but produced by CO_2 flooding (crude oil #1).

crude oil #2 (continuous trace) is produced by primary production (no CO_2 injection). Figure 9.7 shows the LC trace for the asphaltene solubility profile for both tested materials, and the corresponding ΔPS values. The results clearly indicate that the crude oil produced by CO_2 flooding is less stable ($\Delta PS = 3.72$) toward asphaltene precipitation than the crude produced by primary production ($\Delta PS = 1.49$). As mentioned, a $\Delta PS > 2.50$ indicates high asphaltene precipitation risk [27,28]. In this case, CO_2 flooding is decreasing asphaltene stability, and therefore, increasing the tendency of asphaltenes to come out of solution with the concomitant increase in flow assurance issues during production [31].

Finally, an improved method for the separation of asphaltene solubility fractions was also developed in our laboratories [29]. This method has been proven useful for the characterization of virgin [29], thermally [29], and hydroprocessed [32] heavy crude oils and their fractions. In this case, this method is applied to the characterization of crude oil #4 (13°API) and two steam-treated samples of the same material. A continuous steam injection process has been considered as a potential EOR method for this reservoir, and there were concerns that the thermal treatment led to asphaltene chemical changes and potential solid precipitation.

A typical LC trace for the asphaltene solubility method is shown in Figure 9.8 for crude oil #4 (continuous trace). As in the other methods, maltenes (n-heptane solubles) elute first from the column. Then, the mobile phase is changed in successive steps to solvents of increasing solubility parameters: 15% CH_2Cl_2/85% n-C7 ($\delta = 16.0$ MPa$^{0.5}$), 30% CH_2Cl_2/70% n-C7 ($\delta = 16.8$ MPa$^{0.5}$), 100% CH_2Cl_2 ($\delta = 20.2$ MPa$^{0.5}$), and 10% methanol/90% CH_2Cl_2 ($\delta = 21.1$ MPa$^{0.5}$), to separate the asphaltenes into four different solubility fractions [29]. By this method, four different signals are obtained.

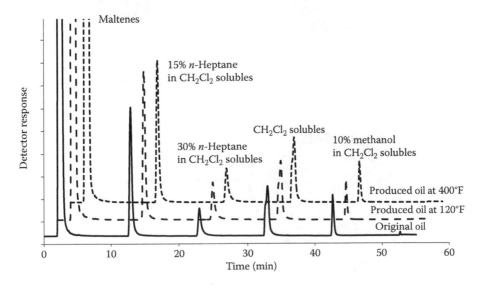

FIGURE 9.8 Comparison of asphaltene solubility fraction analysis of a crude oil and two steam-produced oils.

In Figure 9.8, asphaltene fraction traces for two produced oils from steam treatment tests are shown (discontinuous traces). These two tests were carried out at different temperatures (120°F and 400°F). As the figure shows, no significant changes are observed in the distribution of asphaltenes in the solubility fractions for the two steam-treated samples in comparison with the original crude. Therefore, it can be concluded that the steam process did not affect the asphaltenes present and no asphaltene precipitation issues are expected for these crude oil samples.

9.3.4 ANALYSIS OF DEPOSITS

One of the common operational issues in upstream operation is the formation of solid deposits in subsurface and surface facilities. As shown in Figure 9.9, these solids (Figure 9.9a) or semisolids (Figure 9.9b) are mostly black and are heterogeneous in nature. When one of their main components is asphaltenes, they show high aromaticity and have low H/C molar ratio [4].

The accumulation of these deposits in heat exchangers or other process units leads to plugging in tubes, pumps, valves, and pipelines with significant economic losses throughout the petroleum value chain. Figure 9.10 shows a transversal cut of

(a)   (b)

FIGURE 9.9 Photographs of solid (a) and semisolid (b) deposits from upstream operations.

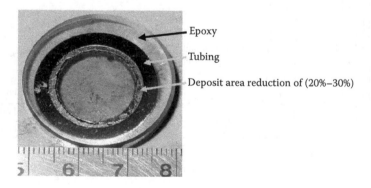

FIGURE 9.10 Photograph of a heat exchanger tube after asphaltene deposition (area reduction of 20%–30%).

a tube from a heat exchanger. As shown, there is deposit formation on the walls of the tube with the concomitant reduction of 20%–30% of the effective area. These results are quite typical not only for upstream but also for refinery units. Therefore, characterization of these materials represents the first step in understanding their formation, precipitation, and accumulation, and in designing mitigation and treatment processes.

With these ideas in mind, three crude oils and their deposits from several Chevron North American operations were analyzed and the results are reported in Table 9.1. As expected, asphaltene concentrations were higher in the deposits (70, 47, and 12 wt.%) than those found in the corresponding crude oils (1.2, 5.6, and 4.2 wt.%, respectively).

Furthermore, when a deposit from production operation was analyzed by using the asphaltene solubility profile method (discontinuous LC trace in Figure 9.11), a vastly different shape is obtained when compared with the original crude oil (continuous trace in Figure 9.11). As shown, the deposit's first peak around 13–14 min

TABLE 9.1

Asphaltene Contents of Produced Crude Oils and Solid Deposits from Upstream Operations

Sample	Asphaltene Content (%)
Crude A (mid-continent)	2.1
Deposit A	70.0
Crude B (Gulf Coast)	5.6
Deposit B	47.1
Crude C (Gulf Coast)	4.2
Crude C	12.1

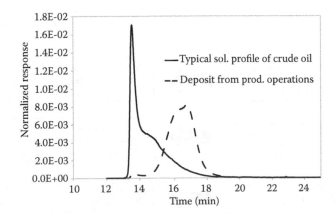

FIGURE 9.11 Comparison between a typical asphaltene solubility profile of a produced crude oil and a deposit from production operations.

decreases in intensity, the solubility distribution is wider, and the second peak shifted toward higher times indicating a higher solubility parameter or lower solubility. In general, similar solubility profiles have been reported previously for processed samples [25,26,33,34].

In the case of processed materials, this type of solubility profile has been attributed to cracking/coking reactions that eliminate alkyl chains and increase the aromatic condensation degree of asphaltenes. These chemical changes decrease the solubility of the asphaltenes, and therefore, shift the second peak to the right of the chromatogram [1,25,26,35–38]. However, at the low temperatures commonly used during upstream operations (<150°C), cracking and aromatic condensation reactions are unlikely. A more plausible explanation for this phenomenon is the preferential precipitation of the lowest soluble molecules present in the produced crudes with the concomitant accumulation to form the solid deposits found in upstream facilities.

9.4 CONCLUSIONS

As summarized in Figure 9.12, the three on-column filtration asphaltene characterization techniques can be successfully used to provide useful information about asphaltene behavior during regular production operations, as well as in the selection and design of treatments and recovery processes. Specifically

- It was found that the on-column filtration technique can be effectively and rapidly used for determining the asphaltene content of crude oil samples from upstream operations.
- The asphaltene solubility profile method can be effectively used to measure the stability of extra heavy oil crude oil/naphtha blends and crude oils during CO_2 flooding, and to screen additives to improve asphaltene solubility.
- Also, the asphaltene solubility fraction method is particularly useful to monitor changes in asphaltene distributions after steam injection processes.
- Finally, using the solubility profile, it was found that deposits from the production operation are preferentially enriched in low-solubility asphaltenes.

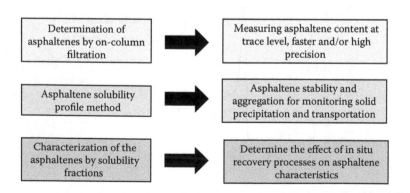

FIGURE 9.12 Application of asphaltene characterization techniques in upstream operations.

ACKNOWLEDGMENT

We thank Chevron Energy Technology Company, in particular the Measurement and Chemistry Focus Area, for providing funding and permission to publish this paper.

REFERENCES

1. Wiehe, I. A., *Chemistry of Petroleum and Macromolecules*. CRC Press, Boca Raton, FL, 2008 and references therein.
2. Kelland, M. A., *Production Chemicals for the Oil and Gas Industry*. CRC Press, Boca Raton, FL, 2009, pp. 111–148 and references therein.
3. Mansoori, G. A., Deposition and fouling of heavy organic oils and other compounds. In *9th International Conference on Properties and Phase Equilibria for Product and Process Design*, Okayama, Japan, 2001.
4. Carbognani, L., Contreras, E., Guimerans, R., León, O., Flores, E., Moya, S., Physicochemical characterization of crudes and solid deposits as a guideline to optimize oil production, SPE 64993. In *2001 SPE International Symposium on Oilfield Chemistry*, Houston, TX, February 13–16, 2001.
5. Abdallah, D., Impact of asphaltenes deposition on completion design for CO_2 pilot in an onshore Abu Dhabi field, SPE-162190. In *Abu Dhabi International Petroleum Exhibition and Conference*, Abu Dhabi, UAE, November 11–14, 2012.
6. Hervey, J. R., Iakovakis, A. C., Performance review of a miscible CO_2 tertiary project: Rangely Weber Sand Unit, Colorado. *SPE Reserv. Eng.*, 6 (2), 163–168, 1991.
7. Beliveau, D., Payne, D. A., Analysis of a tertiary CO_2 flood pilot in a naturally fractured reservoir, Paper SPE 22947. In *SPE Annual Technical Conference and Exhibition*, Dallas, TX, October 6–9, 1991.
8. Peterson, C. A., Pearson, E. J., Chodur, V. T., Periera, C., Beaver Creek Madison CO_2 enhanced recovery project case history; Riverton, WY, Paper SPE 152862. In *SPE Improved Oil Recovery Symposium*, Tulsa, OK, April 14–18, 2012.
9. Turta, A. T., Najman, J., Singhal, A. K., Leggitt, S., Fisher, D., Permeability impairment due to asphaltenes during gas miscible flooding and its mitigation, SPE 37287. In *SPE International Symposium on Oilfield Chemistry*, Texas, 1997.
10. Vargas, F., Development of experimental and modeling methods to understand asphaltene precipitation. In *12th International Conference on Petroleum Phase Behavior and Fouling (Petrophase)*, St. Petersburg Beach, FL, June 10–14, 2012.
11. Ovalles, C., Rogel, E., Segerstrom, J., Improvement of flow properties of heavy oils using asphaltene modifiers, SPE-146775. In *SPE Annual Technical Conference and Exhibition*, Denver, CO, 2011, and references therein.
12. Altgelt, K. H., Harle, O. L., The effect of asphaltenes on asphalt viscosity. *Ind. Eng. Chem. Prod. Res. Dev.*, 14, 242, 1975.
13. Rimmer, D. P., Gregori, A. A., Hamshar, J. A., Yildirim, E., Pipeline emulsion transportation for heavy crude oils. In *Fundamentals and Applications in the Petroleum Industry*, Scharamm, L. L. (Ed.). ACS, Washington, DC, 1992 and references therein.
14. Núñez, G. A., Rivas, H. J., Rodriguez, D. J., Layrisse, I. A., Development of a new technology: Profiting from temporary setbacks during scale-up. *J. Petr. Tech.*, 47, 400, 1995 and references therein.
15. Verzaro, F., Bourrel, M., Garnier, O., Zhou, H., SPE 78959. In *SPE International Thermal Operations and Heavy Oil Symposium and International Horizontal Well Technology Conference*, Calgary, Alberta, Canada, November 4–7, 2002.
16. Argillier, J.-F., Henaut, I., Gateau, P., Heraud, J.-P., In SPE International Thermal Operations and Heavy Oil Symposium Proceedings, 2005, PS2005-343 and references therein.

17. Salager, J.-L., Briceno, M. I., Bracho, C. L., Heavy hydrocarbon emulsions. Making use of the state of the art in formulation engineering. In *Encyclopedic Handbook of Emulsion Technology*, Sjoblom, J. (Ed.). Marcel Dekker, New York, 2001, p. 460.
18. Canadian Patent CA 2019760 901231, assigned to Air Products, 1990.
19. French Patent FR 2480403, assigned to Institute Francais du Petrole, 1981.
20. French Patent FR 2842885, assigned to Institute Francais du Petrole, 2004.
21. French Patent FR 2878937, assigned to Institute Francais du Petrole, 2006.
22. Storm, D. A., US Patent No. US 6,178,980, 2001.
23. Diebold, J. P., Czernik, S., Additives to lower and stabilize the viscosity of pyrolysis oils during storage. *Energy Fuels*, 11, 1081, 1997.
24. Varadaraj, R., Brons, C. H., US Patent Application US 2006/0000749, 2006 and references therein.
25. Schabron, J. F., Rovani, J. F., Sanderson, M. M., Asphaltene determinator method for automated on-column precipitation and redissolution of pericondensed aromatic asphaltene components. *Energy Fuels*, 24, 5984, 2010.
26. Rogel, E., Ovalles, C., Moir, M., Determination of asphaltenes in crude oil and petroleum products by the on column precipitation method. *Energy Fuels*, 23, 4515, 2009.
27. Rogel, E., Ovalles, C., Moir, M. E., Asphaltene stability in crude oils and petroleum materials by solubility profile analysis. *Energy Fuels*, 24 (8), 4369–4374, 2010.
28. Rogel, E., Ovalles, C., Moir, M. E., Asphaltene chemical characterization as a function of solubility: Effects on stability and aggregation. *Energy Fuels*, 26, 2655, 2012.
29. Ovalles, C., Rogel, E., Moir, M. E., Thomas, L., Pradhan, A., Characterization of heavy crude oils, their fractions, and hydrovisbroken products by the asphaltene solubility fraction method. *Energy Fuels*, 26, 549, 2012.
30. Standard test method for determination of asphaltenes (heptane insolubles) in crude petroleum and petroleum products. ASTM 6560. ASTM International USA, 2005.
31. Leontaritis, K. J., Mansoori, G. A., Asphaltene deposition: A survey of field experiences and research approaches. *J. Pet. Sci. Eng.*, 1, 229–239, 1988.
32. Ovalles, C., Rogel, E., Lopez, J., Pradhan, A., Moir, M. E., Predicting reactivity of feedstocks to resid hydroprocessing using asphaltene characteristics. *Energy Fuels*, 27, 6552, 2013.
33. Lopez-Linares, F., Carbognani, L., Hassan, A., Pereira-Almao, P., Rogel, E., Ovalles, C., Pradhan, A., Zintsmaster, J., Adsorption of athabasca vacuum residues and their visbroken products over macroporous solids: Influence of their molecular characteristics. *Energy Fuels*, 25, 4049, 2011.
34. Rogel, E., Ovalles, C., Pradhan, A., Leung, P., Chen, N., Sediment formation in residue hydroconversion processes and its correlation to asphaltene behavior. *Energy Fuels*, 27, 6587, 2013.
35. Ancheyta, J., Centeno, G., Trejo, F., Speight, J. G., Asphaltene characterization as function of time on-stream during hydroprocessing of Maya crude. *Catal. Today*, 109, 162, 2005.
36. Storm, D. A., Decanio, S. J., Edwards, J. C., Sheu, E. Y., Sediment formation during heavy oil upgrading. *Pet. Sci. Technol.*, 15, 77, 1997.
37. Michael, G., Al-Siri, M., Khan, Z. H., Ali, F. A., Differences in average chemical structures of asphaltene fractions separated from feed and product oils of a mild thermal processing reaction. *Energy Fuels*, 19, 1598, 2005.
38. Speight, J. G., New approaches to hydroprocessing. *Catal. Today*, 98, 55, 2004.

10 Asphaltene Adsorption on Iron Oxide Surfaces

Estrella Rogel and Michael Roye

CONTENTS

CONTEXT

- Asphaltene–iron oxide interactions play a significant role in the formation of upstream deposits.
- Adsorbed species on iron oxide are predominantly difficult-to-dissolve asphaltene molecules (DDAs) that form aggregates. This might help to explain the enrichment of deposit in this type of asphaltenes.

ABSTRACT

Asphaltene adsorption on iron oxide (Fe_2O_3) from toluene solutions and crude oils was determined experimentally and correlated with their stability and aggregation behavior. The isotherms of pentane asphaltenes from toluene indicate an adsorption process that can be described using the classic Langmuir isotherm. This adsorption process is endothermic in nature. This means that it is driven by the increase of entropy either in the surface or in the solvent. It was also found that the adsorbed species are predominantly asphaltene molecules that form part of aggregates and can be identified as difficult-to-dissolve asphaltenes (DDAs). The kinetic analysis of asphaltene adsorption from three different crude oils on Fe_2O_3 particles indicates that equilibrium is reached within 24 h. At long contact times (1–72 h), the adsorption kinetics seems to follow a pseudo second order, indicating that in this range, the adsorption is controlled by chemisorption.

10.1   INTRODUCTION

The adsorption of polar crude oil components in different minerals and metal surfaces has been linked to wettability changes [1], permeability losses [2], adhesion of organic deposits on tubing walls [3], etc. In particular, deposition of organic material on the reservoir and process equipment poses production, transport, and refining challenges. Frequently, analysis of asphaltene deposits from different sources (upstream and downstream) reveals the presence of iron compounds as significant components. For instance, in a set of around 50 samples analyzed from Venezuelan oil fields, iron was observed in almost all of them [4,5]. In fact, it has been reported that several costly production problems arise from the interaction between iron oxides and the asphaltene fraction of crude oils [6]. During downstream operations, our own experience in analyzing coke and asphaltene deposits indicates that the presence of iron compounds (iron oxide and iron sulfide) is a common occurrence. Furthermore, there is evidence that the presence of iron in crude oils is associated with a lower stability toward asphaltene precipitation [7]. Screening of the interaction between iron compounds and asphaltenes showed that Fe_2O_3 was the most active among them [8], and more recently, the formation of iron complexes with oxygen functionalities from the crude oil has been reported [6].

The impact of iron and iron compounds on routine operations in the oil industry has been recognized for a long time [9]. Iron has been reported as an important component of solids recovered from wells [3,4]. Furthermore, the stages of deposit formation include steel corrosion products formed in situ and the adsorption or chemisorption of organic compounds on the modified tubing steel surface and mineral pores [10]. Acid stimulation of reservoirs has been observed to cause ferric iron (Fe^{3+}) sludges [11,12]. It is believed that iron increases the amount of polar components precipitated from the crude oil [13]. Studies on fine tailings produced during bitumen extraction have revealed that the yield of unwanted emulsions [14] and the amount of insoluble organics is a function of iron content [15].

Iron compounds have been used as catalysts for hydrogen activation [16,17], carbon–carbon bond scission [18], and oxidation [19]. In fact, a probable catalytic interaction between magnetite and petroleum at well conditions has been studied [3], and the results indicate the formation of complexes between iron ions and organic oxygen functionalities. Selective adsorption of carboxylic compounds has been reported for iron minerals such as goethite [20] and hematite [21].

Theoretical studies of the interactions between asphaltene model molecules and iron indicated the formation of bonds on aromatic rings and directly on heteroatoms, and thus a decrease of C–C, N–C, and S–C bond energies (bond activations) with formation of metal–asphaltene complexes [22]. Other theoretical ab initio calculations of the interaction between asphaltenes and iron oxide indicated that molecules with high aromaticity and low H/C ratio exhibited the highest adsorption energy [23], and attractive interactions are located mainly at the asphaltene aromatic region [24].

Spectroscopy studies of the interactions of iron compounds and asphaltenes indicated that Fe_2O_3 was the most active iron oxide species: Fe_2O_3 seems to affect the H-bonding tendency of asphaltenes and resins, similar to aging [8]. It has also been shown that the insertion of iron in asphaltenes significantly increases their molecular weight, as measured by vapor pressure osmometry [25], an effect that might be related to the enhancement of H-bonding.

Furthermore, asphaltene adsorption has been studied on a variety of absorbents, including clays [26–28], glass and silica [29], pure oxides [26,27,30], and metal surfaces [31–34].

Most of the adsorption behavior observed for asphaltenes has been fitted using the Langmuir equation. However, in a few cases, multilayer adsorption (or in steps) [28,29,35] has been found and linked to asphaltene aggregation and lateral interactions on the adsorbed layer. Higher asphaltene adsorption has also been linked to higher acidity of the adsorbent [36]. In fact, several authors report that polar interactions between asphaltenes and surfaces are responsible for adsorption [26,32]. Increasing the nitrogen content increases the amount of asphaltenes adsorbed [26].

In this work, the adsorption of pentane-extracted asphaltenes (C_5 asphaltenes) in iron oxide Fe_2O_3 is studied to evaluate its effect on asphaltene aggregation and stability. C5 asphaltene solutions were analyzed before and after adsorption using asphaltene solubility profile [37], fluorescence, and size exclusion liquid chromatography. Kinetics of the adsorption from real samples (crude oils) was also determined for three different crude oils. Similar analyses of crude oil were also carried out before and after adsorption.

The main goal of this work is to understand the effect that the presence of minerals or other substrates has on the behavior of asphaltenes and other polar species, and its potential effect not only in upstream processes but throughout the petroleum value chain.

10.2 EXPERIMENTAL

10.2.1 Samples and Materials

Asphaltenes were extracted from a heavy crude oil atmospheric residue using pentane as precipitant agent (C5 asphaltenes) following a modification of ASTM

TABLE 10.1

Elemental Composition and Other Properties of C5 Asphaltenes

	C (wt.%)	H (wt.%)	N (wt.%)	S (wt.%)	Ni (ppm)	V (ppm)	Molecular Weight (g/mol)	Density (g/cm³)
C5 asphaltenes	83.39	8.06	1.76	5.27	368	1674	904	1.155

6560 [38]. Table 10.1 shows the elemental composition and molecular weight of these asphaltenes.

- Three crude oils obtained from the same oilfield in North America: A1, A2, and A3
- Fe_2O_3 (Fisher), characteristics: 100 mesh; surface area 5.57 m²/g, measured using the BET method
- High-performance liquid chromatography–grade toluene, methylene chloride, heptane, and methanol

10.2.2 ADSORPTION ISOTHERM STUDIES OF ASPHALTENE SOLUTIONS

Adsorption isotherms were obtained for C5 asphaltenes at 30°C, 45°C, and 60°C using the following procedure: solutions were prepared in toluene, ranging from 50 mg/L to 30 g/L. The prepared solutions (5.0 mL) were mixed with 0.5 g iron oxide. The solutions were incubated for 24 h at their predetermined temperatures in a shaker. After 24 h, the solutions were analyzed to determine the asphaltene concentration, stability by using the asphaltene solubility profile [37], aggregation behavior by using fluorescence, and size distribution by using size exclusion liquid chromatography (the details of these techniques will be presented in Sections 10.2.6 and 10.2.7, respectively).

10.2.3 ADSORPTION KINETICS OF ASPHALTENES FROM CRUDE OILS

The adsorption kinetics of asphaltenes was studied for three crude oils, A1, A2, and A3, from an oilfield in North America. The crude oil samples were agitated with iron oxide at a constant temperature of 30°C in an incubator. At different times, ranging from 5 min to 72 h, the solutions were taken and analyzed to determine the asphaltene concentration, stability, and distribution by using the asphaltene solubility profile.

10.2.4 DETERMINATION OF ASPHALTENE CONCENTRATION

Asphaltene concentrations were determined using an in-house method known as the on-column filtration technique [39]. In this technique, a solution of the sample is injected in a column packed with an inert material, using *n*-heptane as the mobile phase. This solvent induces the precipitation of asphaltenes and, as a consequence, their retention in the column. The maltenes are eluted first. After the maltenes have eluted, the mobile phase is switched to a blend of dichloromethane/methanol 90:10 v/v, which redissolves the asphaltenes. Asphaltenes are quantified using an evaporative light scattering detector (ELSD) Alltech 2000. A calibration using extracted asphaltenes as standards was used in the determination of the concentration.

10.2.5 EVALUATION OF THE STABILITY AND ADSORBED SPECIES USING ASPHALTENE SOLUBILITY PROFILE

The solubility distribution of the solutions was determined using a technique based on on-column filtration [36,40]. In this technique, a solution of the sample in methylene chloride is injected in a column packed with an inert material using n-heptane as the mobile phase. This solvent induces the precipitation of asphaltenes and, as a consequence, their retention in the column. The first eluted fraction from the column is the maltenes, which are soluble in n-heptane. After all of this fraction has eluted, the mobile phase is changed gradually from pure n-heptane to 90:10 methylene chloride/methanol and then to 100% methanol.

This procedure gradually redissolves the asphaltenes from the easy to dissolve (low solubility parameter) to the difficult to dissolve (high solubility parameter). Asphaltenes can be quantified using an ELSD Alltech 2000.

It has been found [36,40] that asphaltene species that elute first are considered easy-to-dissolve asphaltenes (EDAs), whereas the species with higher retention times are named difficult-to-dissolve asphaltenes (DDAs). Similarly, for this technique, a parameter, ΔPS, is used to quantify stability. In general, a larger ΔPS indicates lower stability and a higher tendency toward asphaltene precipitation [36,40].

10.2.6 FLUORESCENCE MEASUREMENTS

Fluorescence measurements were performed for the low-concentration solutions before and after adsorption (around 50–300 ppm) using a fluorescence spectrometer Hitachi Model F-4500 with a 150-W xenon lamp as the excitation source. The excitation wavelength was 310 nm, and fluorescence was determined in the range 200–900 nm. A 10-mm path-length cell was used during the measurements. The goal of the fluorescence measurements was to qualitatively compare the nature of the asphaltenes that remained in solution after adsorption with those present in the original sample.

10.2.7 SIZE EXCLUSION CHROMATOGRAPHY

Size exclusion chromatography (SEC) studies were done using a 30 cm × 0.10 cm Mixed E column with a blend 90:10 methylene chloride/methanol. The flow rate was 1.0 mL/min. The temperature was kept constant at 25°C. The injection volume was 80 μL, and the goal was again to compare the solutions before and after adsorption. ELSD Alltech 2000 was used in this set of experiments.

10.3 RESULTS AND DISCUSSION

10.3.1 C5 ASPHALTENE ADSORPTION

10.3.1.1 Adsorption Isotherm Analysis

Adsorption isotherms for C5 asphaltenes are shown at three different temperatures in Figure 10.1. For the two lowest temperatures (30°C and 45°C), adsorption isotherms are practically identical and the results are within the error of determination. However, when the temperature increases to 60°C, there is a significant decrease in the adsorbed amount. A previous study [31] of asphaltene adsorption on stainless steel (304L) reported similar adsorption behavior; that is, at room or slightly higher temperatures, adsorption isotherms look similar, while at 60°C the decrease in adsorption is noticeable. This behavior is due to the increase in the solubility of this fraction as the temperature increases. This makes asphaltenes more prone to remain in solution.

Isotherms presented in Figure 10.1 appear to be of the Langmuir type, so they can be fitted using the Langmuir equation:

$$\frac{C_{eq}}{\Gamma_{eq}} = \frac{C_{eq}}{\Gamma_{max}} + \frac{1}{K_{eq}\Gamma_{max}} \tag{10.1}$$

where C_{eq} is the equilibrium asphaltene concentration, Γ_{eq} is the equilibrium asphaltene adsorbed amount, Γ_{max} is the maximum asphaltene adsorbed amount, and K_{eq} is the equilibrium constant.

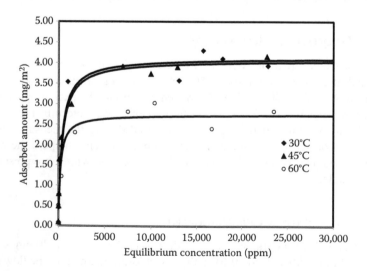

FIGURE 10.1 Adsorption isotherms of C5 asphaltenes on iron oxide at different temperatures.

TABLE 10.2

Langmuir Fitted Values of Γ_{max} and K_{eq} at Different Temperatures for the Adsorption of C5 Asphaltenes on Iron Oxide

Temperature (°C)	Maximum Adsorbed Amount Γ_{max} (mg/m²)	Equilibrium Constant (L/mg)	Regression Coefficient
30	4.07	0.0025	0.9943
45	4.13	0.0027	0.9980
60	2.74	0.0047	0.9888

Values of Γ_{max} and K_{eq} at different temperatures are shown in Table 10.2. In terms of the maximum adsorbed amounts, the values obtained are on the same order of magnitude reported previously for asphaltene adsorption on metals and metal oxides [31–34].

As mentioned before, the shape of the isotherms shown in Figure 10.1 is characteristic of Langmuir adsorption isotherms. However, this does not mean that the assumptions behind the Langmuir equation are fulfilled by the asphaltene adsorption. These assumptions are as follows: formation of a chemisorbed monolayer on a homogeneous surface of identical sites that are equally available and energetically equivalent such that each site carries an equal number of molecules that do not interact with one another (no adsorbate–adsorbate interactions) [41]. In fact, several authors have concluded that asphaltenes adsorb, forming multilayers [28,29,35]. A simple calculation of the number of molecules per area using the data in Table 10.2 and a molecular weight of 904 g/mol (obtained based on SEC experiments) yields values from 3 molecules/nm² at 30°C and 45°C to 2 molecules/nm² at 60°C. These values indicate a multilayer adsorption and are within the range of the value reported for Fe_2O_3 nanoparticles at 25°C [30]. Additionally, an estimation of the average molecular area of the C5 asphaltene using the experimental density (1.155 g/cm³), assuming a cylindrical shape for asphaltenes and a parallel configuration of the polyaromatic rings with respect to the surface, indicates a multilayer structure of 6–10 stacked molecules at the maximum coverage.

Other isotherm equations [42], Freundlich, Temkin, and P–R, were also used to fit the experimental adsorption data; however, the Langmuir plots have the best fitting. The worst fitting corresponds to Freundlich equation. Interestingly, the Freundlich equation applies to multilayer adsorption.

10.3.1.2 Effect of Temperature and Thermodynamic Parameters

The Gibbs free energy of adsorption ΔG_{ads} can be calculated using the following equation:

$$\Delta G_{ads} = -RT\ln K \qquad (10.2)$$

where ΔG_{ads} is the free energy of adsorption, R is the universal gas constant, T is the temperature, and K is the adsorption equilibrium constant determined using the Langmuir equation and the asphaltene molecular weight.

TABLE 10.3

Free Energy of Adsorption (ΔG) for C5 Asphaltenes on Iron Oxide as a Function of Temperature

Temperature (°C)	ΔG (kJ/mol)
30	−25.1
45	−26.5
60	−29.2

Values for the Gibbs free energy of adsorption are reported in Table 10.3 as a function of temperature. Similar values have been reported previously by other authors for asphaltene adsorption on metal surfaces and metal oxides [30,33]. These values indicate that the adsorption of asphaltenes is a spontaneous process at the studied temperatures.

Enthalpy and entropy for the adsorption process can be obtained by using the equation

$$\ln K = \frac{\Delta S}{R} - \frac{\Delta H}{RT} \tag{10.3}$$

ΔH and ΔS are obtained from the slope and intercept of van't Hoff plot of $\ln K$ versus $1/T$. In this case, ΔH (16.5 kJ/mol) and ΔS (0.13 kJ/K mol) are both positive. Therefore, on the basis of the equilibrium constants obtained using the Langmuir equation, the adsorption process is endothermic in nature, suggesting that the process is driven by the increase of entropy either in the surface or in the solvent. These results should be taken cautiously since it seems that asphaltene adsorption does not fulfill the assumptions made by the Langmuir equation. Additionally, asphaltenes are a complex mixture of different components, not a single compound. However, it is important to state that the previously mentioned isotherm equations also indicated that asphaltene adsorption is an endothermic process.

10.3.1.3 Adsorbed Species

There have been several studies analyzing the species that preferentially adsorbed from asphaltene solutions onto minerals and metals. Iron oxides and hydroxides are prone to form stable bonds with a variety of functionalities containing oxygen ($-COOH$, $=CO$, $-OH$), nitrogen ($-NH_2$), and sulfur ($-SH$) or mixed functionalities. X-ray photoelectron spectroscopy analysis of adsorbed asphaltenes revealed the presence of carboxylic, thiophenic, sulfide, sulfoxide, pyridinic, and pyrrolic-type functional groups, similar to the functionalities present in bulk asphaltenes [33,43].

In the present work, the characteristics of the adsorbed species were evaluated by examining the asphaltenes remaining in solution after the adsorption took place. Three different techniques were used: solubility profile, SEC, and fluorescence measurements.

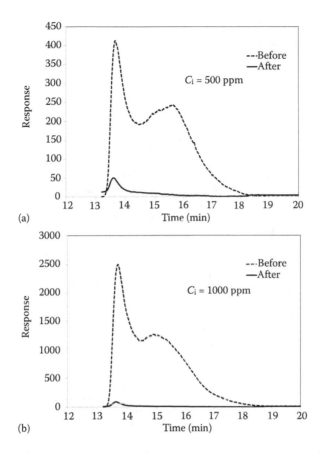

FIGURE 10.2 Solubility profile of solution before and after C5 asphaltene adsorption at $T = 45°C$. Initial concentrations: (a) $C_i = 500$ ppm; (b) $C_i = 1000$ ppm.

10.3.1.3.1 Asphaltene Solubility Profile Analysis

Solubility profile analysis was used to examine the changes in solubility properties of the asphaltenes before and after adsorption. As shown in Figures 10.2 and 10.3, molecules from the whole range of solubility are adsorbed. However, adsorption of the DDAs dominates. This is particularly clear at low concentrations, as can be seen in Figure 10.2. However, easy-to-dissolve molecules also adsorb and it seems that as a consequence of this adsorption, the stability of the asphaltenes in solution decreases slightly at high concentrations, as shown by the increase in the stability parameter ΔPS (Figure 10.4) [39]. Results shown in Figures 10.2 through 10.4 correspond to the adsorption at 45°C. Similar results were obtained at the other two studied temperatures.

10.3.1.3.2 Size Exclusion Chromatography

SEC was used to compare solutions before and after adsorption in terms of their apparent asphaltene molecular weights. The results indicate that at low concentrations,

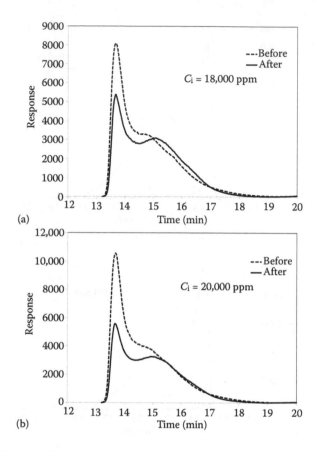

(a)

(b)

FIGURE 10.3 Solubility profile of solution before and after C5 asphaltene adsorption at $T = 45°C$. Initial concentrations: (a) $C_i = 18,000$ ppm; (b) $C_i = 20,000$ ppm.

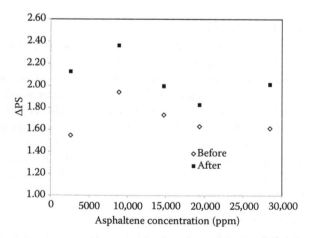

FIGURE 10.4 Stability of the C5 asphaltene solutions toward precipitation before and after adsorption on iron oxide at $T = 45°C$.

the predominant adsorbed species are those molecules that form aggregates or have the largest molecular weights. An example of this behavior is shown in Figure 10.5 for two of the lowest initial concentrations. On the basis of the results from Section 10.3.1.3.1, it seems that the molecules that form aggregates or with larger molecular weight also exhibit low solubility (DDAs). This is expected as the formation of aggregates can be considered as the formation of a second phase due to low solubility.

It is also interesting to note that both examples in Figure 10.5 show a larger concentration of the lowest molecular weight molecules (right of the chromatograph), indicating the possibility that the break of aggregates might free some small trapped molecules.

At higher concentrations in the plateau region of the isotherm (see Figure 10.6), the comparison still showed a smaller concentration of molecules in the higher

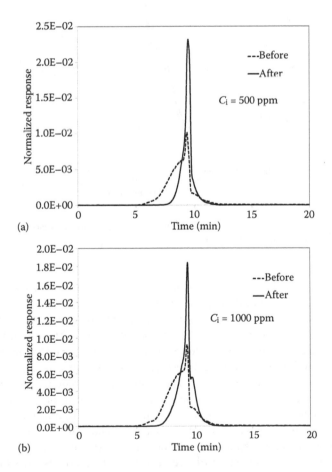

(a)

(b)

FIGURE 10.5 SEC chromatograph before and after C5 asphaltene adsorption at $T = 45°C$. Initial concentrations: (a) $C_i = 500$ ppm; (b) $C_i = 1000$ ppm.

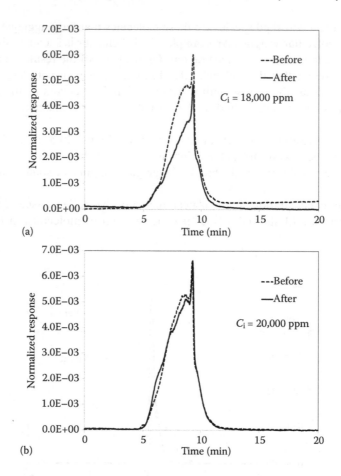

FIGURE 10.6 SEC chromatograph before and after C5 asphaltene adsorption at $T = 45°C$. Initial concentrations: (a) $C_i = 18,000$ ppm; (b) $C_i = 20,000$ ppm.

molecular weight region of the chromatographs; however, this difference becomes less noticeable as the concentration increases.

On the basis of the results, it can be said that at low concentrations, the molecules forming aggregates are adsorbed preferentially and, therefore, the solution becomes depleted of aggregates. However, the disappearance of aggregates can be explained by just the decrease in concentration because of adsorption. As a consequence of this decrease, the concentration of asphaltenes is below a certain threshold (equivalent to the critical micelle concentration for surfactants in aqueous solutions) at which the aggregates start to form. To check this possibility, comparisons were made between solutions before and after adsorption that have approximately the same asphaltene concentrations. Two examples at low asphaltene concentrations are shown in Figure 10.7. In both examples, two solutions are compared that have approximately the same concentration. The main difference is that one of the solutions has been in contact with iron oxide (after) and adsorption

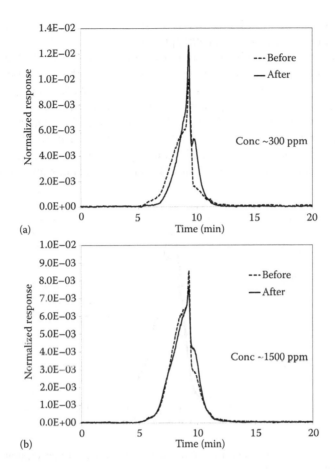

FIGURE 10.7 SEC chromatographs of solutions with similar concentrations before and after C5 asphaltene adsorption at $T = 45°C$. (a) Concentration ~300 ppm; (b) concentration ~1500 ppm.

has taken place while the other solution has not adsorbed (before). These examples indicate that there is a decrease of molecules in the high molecular weight area and an increase in the low molecular weight area when solutions with similar concentrations before and after adsorption are compared. This is an indication that the disaggregation is induced by the preferential adsorption of some molecules. At the same time, the disaggregation frees small molecules that were previously part of the aggregates.

The differences observed in Figure 10.7 became less significant, however, as the concentration increases at the plateau region of the isotherms.

10.3.1.3.3 Fluorescence Measurements

It is known that molecular aggregation is accompanied by a substantial decrease in fluorescence efficiency [44]. In fact, integrated areas of fluorescence peaks can decrease by as much as 94% [45]. For this reason, fluorescence spectroscopy has been used to study aggregation processes in asphaltene solutions at relative low concentration ranges (between 25 and 500 ppm) [46–48].

Asphaltene solutions with concentrations <1000 ppm were evaluated before and after adsorption to examine changes upon adsorption. At the lowest initial concentration (Figure 10.8), a decrease in intensity was observed after adsorption, indicating that there are less chromophores available for fluorescence. This is a logical consequence of adsorption as there are fewer molecules in solution. However, at larger concentrations, an increase in intensity is observed in the fluorescence of the solutions after adsorption, indicating that aggregates break down as molecules are adsorbed onto the iron oxide surface. Figure 10.9 contains an example of this behavior.

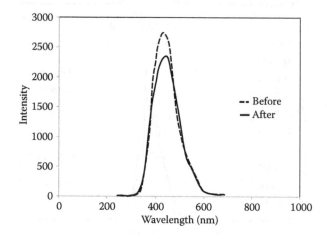

FIGURE 10.8   Fluorescence spectra before and after C5 asphaltene adsorption at $T = 45°C$. Initial concentration: 50 ppm.

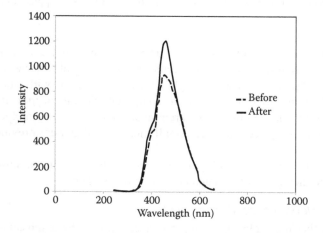

FIGURE 10.9   Fluorescence spectra before and after C5 asphaltene adsorption at $T = 45°C$. Initial concentration: 500 ppm.

10.3.2 CRUDE OILS

Samples of different crude oils were examined to evaluate the adsorption of asphaltenes when the crude oils come into contact with iron oxide, which is a common component in reservoirs and is present in pipelines and other equipment.

10.3.2.1 Kinetic Analysis of Asphaltene Adsorption

Figure 10.10 shows adsorption kinetics at 30°C of asphaltenes from A1, A2, and A3 crude oils. As can be seen, the adsorption rate is significantly high during the first 10 h, decreasing to a plateau at around 20 h. After this time, no significant increase in adsorption was observed up to 72 h, which was the largest time studied.

There is a significant number of models reported in the literature that attempt to quantitatively describe the kinetic behavior during the adsorption process [49]. Each model has a series of limitations based on the specific assumptions made for its development [50]. Additionally, most of the models were built for aqueous solutions and do not take into consideration the possible aggregation of adsorbate molecules onto the solid surface.

However, before applying different models to the set of data shown in Figure 10.10, it could be convenient to examine the different steps that control the adsorption rate for dyes [49] and compare them to the asphaltene case:

a. Diffusion from the bulk solution to a film: It has been shown in Sections 10.3.1.3.2 and 10.3.1.3.3 that asphaltene adsorption at low concentrations involves the destruction of aggregates and release of small molecular weight molecules. Therefore, it seems likely that asphaltenes diffuse as monomers.
b. Diffusion from the film to particle surface, i.e., "film diffusion."

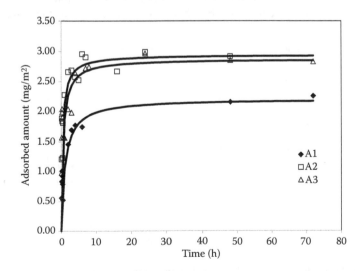

FIGURE 10.10 Adsorption kinetics of asphaltenes from different crude oils.

   c. Migration inside the adsorbent particle by "surface diffusion" or diffusion within liquid-filled pores, i.e., "pore diffusion."

   d. Uptake, which includes several ways of interaction such as chemisorption, physisorption, ion exchange, or complexation. In the case of asphaltenes, this could include asphaltene–asphaltene interactions to form surface aggregates.

The kinetic models used in studying the mechanism of adsorption include pseudo-first-order, second-order, external diffusion, film diffusion, internal diffusion, and the Elovich equation [51]. To understand the predominant mechanism as a function of time, the plot in Figure 10.10 was divided into the following time regions:

   a. 0–30 min: At short times, the data seem to indicate that after a fast adsorption process, some desorption occurs followed by adsorption again. Figure 10.11 shows this behavior for the studied crude oils. Several hypotheses can explain this behavior: (i) at short times, larger errors in the data are expected, and therefore, the behavior observed is product of the scattering of the data; (ii) the presence of crude oil induces changes in the iron oxide that contribute to desorption, i.e., it is possible that some asphaltene molecules induce dispersion of iron oxide particles into the liquid causing the apparent desorption of asphaltenes from the surface; (iii) the adsorption of asphaltenes can be followed by changes in the configuration of molecules in the interface, followed by the release of some of them to the bulk of the solution. There were not enough data points and the scattering was too high to attempt a fitting in this region. On the other hand, data that will be presented in Section 10.3.2.2 supports the third hypothesis.

   b. 1–72 h: At large times, the best fitting was obtained by an equation that describes a pseudo-second-order adsorption, indicating that, in this range, the adsorption is controlled by chemisorption, either by the interaction

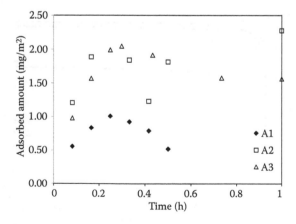

FIGURE 10.11 Adsorption kinetics of asphaltenes from different crude oils at short contact times.

between asphaltenes and iron oxide or by interaction among adsorbed asphaltenes and bulk asphaltenes. In this case, the adsorption rate can be expressed as [51]

$$\frac{d\Gamma_t}{dt} = k(\Gamma_{max} - \Gamma_t)^2 \tag{10.4}$$

where Γ_t and Γ_{max} represent the amount adsorbed at time t and the amount adsorbed at equilibrium, respectively; k is the rate constant of pseudo-second-order adsorption ($m^2\ mg^{-1}\ h^{-1}$). For the boundary conditions $t = 0$ to $t = t$ and $\Gamma_t = 0$ to $\Gamma_t = \Gamma_t$, the integrated form of Equation 10.4 becomes

$$\frac{1}{(\Gamma_{max} - \Gamma_t)} = \frac{1}{\Gamma_t} + kt \tag{10.5}$$

Equation 10.5 can be rearranged to obtain

$$\frac{t}{\Gamma_t} = \frac{1}{k\Gamma_{max}^2} + \frac{1}{\Gamma_{max}}t \tag{10.6}$$

where h is defined as $k\Gamma_{max}^2$, the initial adsorption rate as $\frac{\Gamma_t}{t} \to 0$. Then, Equation 10.6 becomes

$$\frac{t}{\Gamma_t} = \frac{1}{h} + \frac{1}{\Gamma_{max}}t \tag{10.7}$$

Table 10.4 shows the initial adsorption rate as well as the rate constant for each of the crude oils examined. Figure 10.12 show the pseudo-second-order adsorption kinetics for the studied crude oils.

TABLE 10.4
Pseudo-Second-Order Adsorption Kinetics Fitted Parameters for the Adsorption of Asphaltenes from Crude Oils on Iron Oxide

Crude Oil	Maximum Adsorbed Amount Γ_{max} (mg/m²)	Initial Adsorption Rate (mg m⁻² h⁻¹)	Rate Constant (m² mg⁻¹ h⁻¹)	Regression Coefficient
A1	2.28	1.64	0.317	0.9994
A2	2.94	7.61	0.881	0.9986
A3	2.87	5.41	0.655	0.9994

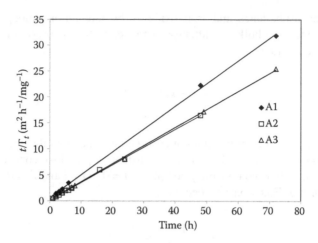

FIGURE 10.12   Pseudo-second-order adsorption kinetics for the studied crude oils.

10.3.2.2   Adsorbed Species

The adsorbed species were evaluated by examining the asphaltenes remaining in solution after the adsorption took place. Only the solubility profile was used since SEC and fluorescence measurements will require preparative separation of the asphaltenes for the analysis.

10.3.2.2.1   Solubility Profile Analysis

To evaluate the characteristics of the adsorbed species, it is convenient to look at time intervals, as was done in Section 10.3.2.1:

a. 0–30 min: Data seem to indicate that after a fast adsorption process, some desorption occurs followed by adsorption again. During this period, the solubility profile measurements revealed that at the beginning of the adsorption, the amount of EDAs in the crude oils decreases, and increases later on. On the other hand, the amount of DDAs seems to decrease slowly over time. Examples of this behavior are shown in Figure 10.13 for crude oils A1 and A2.

These results point toward a mechanism where the EDA molecules adsorb first, probably because they can diffuse faster, and later they are replaced by DDA molecules that are less soluble but move slower toward the surface. This interplay at short contact times should be tested using other crude oils.

b. 1–72 h: Small changes are observed as can be seen in Figure 10.14 for crude oil A3. As was the case in the adsorption of C5 asphaltenes, it was observed that at large contact times, the crude oil becomes slightly less stable.

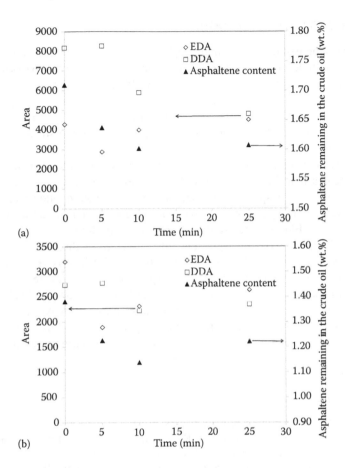

FIGURE 10.13 EDA and DDA areas, and total asphaltene content in the crude oil as a function of contact time. (a) Crude oil A1; (b) crude oil A2.

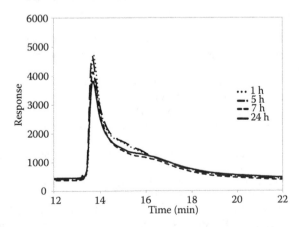

FIGURE 10.14 Changes in the asphaltene solubility profile of crude oil A3 after large contact times with iron oxide.

10.4 CONCLUDING REMARKS

- Adsorption isotherms of pentane asphaltenes from toluene solutions on iron oxide particles appear to be of the Langmuir type. Attempts to describe the isotherms using the Freundlich equation yielded low regression coefficients.
- The results indicate that the adsorption process of pentane asphaltenes from toluene over iron oxide is endothermic in nature. This means that the process is driven by the increase of entropy either in the surface or in the solvent.
- At low concentrations, the adsorbed species are predominantly those molecules that form aggregates. An increase in the concentration of low molecular weight species in the solution after adsorption reflects aggregates breaking apart, releasing low molecular weight molecules trapped on those aggregates. Fluorescence studies support these findings.
- The kinetic analysis of asphaltene adsorption from three different crude oils on iron oxide particles indicates that equilibrium is reached within 24 h. At short times (<1 h), the adsorption kinetics indicates a behavior of adsorption–desorption that involves the fast adsorption and subsequent desorption of the EDA molecules, while adsorption of DDAs takes place at a slower rate. This behavior might help explain the enrichment of deposits in the DDAs. At longer times (1–72 h), the adsorption kinetics seems to follow a pseudo-second-order equation.

REFERENCES

1. Neihof, R. *J. Colloid Interface Sci.* **30**: 128, 1969.
2. Turta, A. T.; Najman, J.; Singhal, A. K.; Leggitt, S.; Fisher, D. 1997 SPE International Symposium on Oilfield Chemistry, Texas, *SPE 37287*, 1997.
3. Cosultchi, A.; Ascensio-Gutierrez, J. A.; Reguera, E.; Zeifert, B.; Yee-Madeira, H. *Energy Fuels* **20**: 1281, 2006.
4. Carbognani, L.; Espidel, Y.; Izquierdo, A. 1st SPE Brazil Sect. et al. International Symposium on Colloid Chemistry in Oil Production: Asphaltenes and Wax Deposition, Rio de Janeiro, Brazil, 1995.
5. Carbognani, L.; Orea, M.; Fonseca, M. *Energy Fuels* **13**: 351, 1999.
6. Taft, C. A.; Sigaud, X.; Almeida, A. L.; Guimaraes, T. C. World Petroleum Congress Proceedings, vol. 3, p. 259, 2002.
7. Hussam, H. I.; Raphael, O. *Energy Fuels* **18**: 1354, 2004.
8. Carbognani, L. *Pet. Sci. Technol.* **18**: 335, 2000.
9. Patton, C. C.; Jessen, F. W. *SPE 1088*, 1964.
10. Cosultchi, A.; Garcia-Figueroa, E.; Mar, B.; Gracia-Borquez, A.; Lara, V. H.; Bosch, P. *Fuel* **81**: 413, 2002.
11. Suzuki, F. Western Regional Meeting, Anchorage, AL, *SPE 26036*, 1993.
12. Fredd, C. N.; Fogler, H. S. *SPE J.* **3**: 34, 1998.
13. Nalwaya, V.; Veerapat, T.; Piumsomboon, P.; Fogler, S. *Ind. Eng. Chem. Res.* **38**: 964, 1999.
14. Majid, A.; Sparks, B. D. *Fuel* **75**: 879, 1996.
15. Kotyar, L. S.; Sparks, B. D.; Kodama, H.; Grattam-Bellew, P. E. *Energy Fuels* **2**: 589, 1988.
16. Ovalles, C.; Filgueiras, E.; Morales, A.; Scott, C. E.; Gonzalez-Gimenez, F.; Embaid, B. P. *Fuel* **82**: 88, 2003.

17. Rosa-Brussin, M. F. *Catal. Rev.* **37**: 1, 1995.
18. Linehan, J. C.; Matson, D. W.; Darab, J. G. *Energy Fuels* **8**: 56, 1994.
19. Nassar, N. N.; Hassan, A.; Carbognani, L.; Lopez-Linares, F.; Pereira-Almao, P. *Fuel* **95**: 257, 2012.
20. Jeon, Y. W.; Yi, S.; Choi, S. J. *Fuel Sci. Technol. Int.* **13**: 195, 1995.
21. Buckley, J. S.; Liu, Y.; Monsterleet, S. *SPE J.* **3**: 54, 1998.
22. Rosales, S.; Machin, I.; Sanchez, M.; Rivas, G.; Ruette, F. *J. Mol. Catal. A Chem.* **246**: 146, 2006.
23. Murgich, J.; Isea, R.; Rogel, E.; Leon, O. *Pet. Sci. Technol.* **19**: 437, 2001.
24. Alvarez-Ramirez, F.; Garcia-Cruz, I.; Tavizon, G.; Martinez-Magadan, J. M. *Pet. Sci. Technol.* **22**: 915, 2004.
25. Pereira, J. C.; Luis, M. A.; Cubillos, P. *Pet. Sci. Technol.* **26**: 181, 2008.
26. Dudasova, D.; Simon, S.; Hemmingsen, P. V.; Sjoblom, J. *Colloids Surf. A Physicochem. Eng. Aspects* **317**: 1, 2008.
27. Dudásováa, D.; Flåtenb, G. R.; Sjöblom, J.; Øyea, G. *Colloids Surf. A Physicochem. Eng. Aspects* **335**: 62, 2009.
28. Marczewski, A. W.; Szymula, M. *Colloids Surf. A Physicochem. Eng. Aspects* **208**: 259, 2002.
29. Acevedo, S.; Ranaudo, M. A.; Garcia, C.; Castillo, J.; Fernandez, A. *Energy Fuels* **17**: 257, 2002.
30. Hosseinpour, N.; Khodadadi, A. A.; Bahramian, A.; Mortazavi, Y. *Langmuir* **29**: 14135, 2013.
31. Alboudwarej, H.; Pole, D.; Svrcek, W. Y.; Yarranton, H. W. *Ind. Eng. Chem. Res.* **44**: 5585, 2005.
32. Xie, K.; Karan, K. *Energy Fuels* **19**: 1252, 2005.
33. Rudrake, A.; Karan, K.; Horton, J. H. *J. Colloid Interface Sci.* **332**: 22, 2009.
34. Balabin, R. M.; Syunyaev, R. Z.; Schmid, T.; Stadler, J.; Lomakina, E. I.; Zenobi, R. *Energy Fuels* **25**: 189, 2011.
35. Acevedo, S.; Castillo, J.; Fernandez, A.; Goncalves, S.; Ranaudo, M. A. *Energy Fuels* **12**: 386, 1998.
36. Nassar, N. N.; Hassan, A.; Pereira-Almao, P. *J. Colloid Interface Sci.* **360**: 233, 2011.
37. Rogel, E.; Ovalles, C.; Moir, M. *Energy Fuels* **24**: 4369, 2010.
38. *Standard Test Method for Determination of Asphaltenes (Heptane Insolubles) in Crude Petroleum and Petroleum Products* ASTM 6560. ASTM International USA, Philadelphia, PA, 2005.
39. Rogel, E.; Ovalles, C.; Moir, M. E.; Schabron, J. F. *Energy Fuels* **23**: 4515, 2009.
40. Rogel, E.; Ovalles, C.; Moir, M. *Energy Fuels* **26**: 2655, 2012.
41. El, Q.; Emad, N.; Allen, S. J.; Walker, G. M. *Chem. Eng. J.* **124**: 103, 2006.
42. Ho, Y. S.; Porter, J. F.; McKay, G. *Water Air Soil Pollut.* **141**: 1, 2002.
43. Abdallah, W. A.; Taylor, S. D. *J. Phys. Chem. C* **112**: 18963, 2008.
44. Langhals, H.; Ismael, R.; Yuruk, O. *Tetrahedron* **56**: 5435, 2000.
45. Ebeid, E. M.; El-Daly, S. A.; Langhals, H. *J. Phys. Chem.* **92**: 4565, 1988.
46. Goncalves, S.; Castillo, J.; Fernandez, A.; Hung, J. *Fuel* **83**: 1823, 2004.
47. Ghosh, A. K.; Srivastava, S. K.; Bagchi, S. *Fuel* **86**: 2528, 2007.
48. Wang, Z.; Li, L.; Shui, H.; Wang, Z.; Cui, X.; Ren, S.; Lei, Z.; Kang, S. *Fuel* **90**: 305, 2011.
49. Khraisheh, M. A. M.; Al-Degs, Y. S.; Allen, S. J.; Ahmad, M. N. *Ind. Eng. Chem. Res.* **41**: 1651, 2002.
50. Weber, W. J.; DiGiano, F. A. *Process Dynamics in Environmental Systems.* Environmental Science and Technology Series. Wiley & Sons: New York, 1996.
51. Ho, Y. S.; McKay, G. *Trans. IChemE* **76B**: 332, 1998.

11 Determination of Asphaltenes Using Microfluidics

Farshid Mostowfi and Vincent Sieben

CONTENTS

CONTEXT

- New methodology for the determination of asphaltene content can substantially save time and cost
- Potential applications throughout the petroleum value chain

ABSTRACT

In this chapter, we will discuss a new approach for performing analytical measurements that could reform the way samples are analyzed in the oil field industry. The utility of microfluidics is demonstrated by implementing a miniaturized version of conventional asphaltene content measurement. Relevant details are discussed regarding the origin of color in crude oils, principles of spectroscopy, the conventional wet chemistry procedure, and initial proof-of-concept experiments. The chapter also explores the principles of fluid mechanics and the operations of the microfluidic system. At the end, we summarize the results from analyzing 52 oils and highlight the improvements of the microfluidic method. We show a strong correlation between the optical spectrum acquired by this technique and the asphaltene content using the wet chemistry method. We believe that by incorporating this technology into the industry, valuable information and data can be generated at a much faster pace and with a higher quality.

11.1   INTRODUCTION

Asphaltenes are alkylated polycyclic aromatic species containing heteroatoms such as sulfur and nitrogen, as well as metals such as vanadium and nickel [1]. They are the heaviest and most polar fraction of a crude oil. Mass spectral and diffusion measurements of asphaltenes point to a molecular weight between 500 and 1000 Da, which is compatible with an alkylated polycyclic structure with roughly seven rings [2–4]. This structure is called the "island" architecture. There is, however, another school of thought that advocates the "archipelago" architecture. In that picture, asphaltenes are composed of alkyl-bridged aromatic and cycloalkyl groups connected with alkyl chains [5,6]. Regardless of their exact molecular structure, asphaltenes can form aggregates primarily through π–π stacking of aromatic rings and other interactions. They commonly form nanoaggregates with an average size between 2 and 20 nm [1,7]. These aggregates are porous; thus, smaller molecules and other mobile species can enter their fractal structure. Asphaltenes tend to deposit onto surfaces. Silica and alumina are the prime examples of almost irreversible adhesion of asphaltenes to solid surfaces, which is why conventional chromatography of asphaltenes is difficult. Metal surfaces such as the inner surface of pipelines are also not immune to the adhesion of asphaltenes. They also adsorb to the oil–water interface, which results in stabilization of emulsion films [8,9]. Asphaltenes are generally considered a nuisance in the industry, with limited commercial value as, e.g., road pavement material. However, they are important in understanding the geochemical background of reservoir, the reservoir connectivity, and the migration of the fluid [3,10].

Perhaps the most important property of asphaltenes relate to their deposition and precipitation tendency. Asphaltenes may precipitate and deposit in the formation, wellbore, pipelines, and refineries. The cost of remediation is often staggering. Asphaltenes precipitate owing to changes in pressure, temperature, or composition of the crude oil. For instance, they can precipitate during the primary depletion because of pressure drop near the wellbore, and cause impairment. They can also precipitate because of compositional changes caused by gas injection in enhanced oil recovery applications. When two zones are produced at the same time, asphaltenes could become unstable due to the comingling of incompatible fluids.

Arguably, asphaltenes are the most controversial chemical fraction of a crude oil. After decades of research, their molecular structure, aggregation kinetics, and interactions with other species in the crude oil are not yet well understood [11–24]. Conventional analytical techniques such as chromatography and mass spectrometry are not capable of fully resolving the asphaltene fraction owing to their size and association [8]. The controversy, in part, stems from the unorthodox definition of asphaltenes. The definition of asphaltenes is based on their solubility rather than the molecular structure. Asphaltenes are generally defined as the fraction of crude oil that is insoluble in a normal alkane such as n-heptane at 1:40 v/v ratio and soluble in aromatic solvents such as toluene. Therefore, asphaltenes have an operational or procedural definition, which leaves a lot of room for ambiguity and confusion. This definition, for instance, completely ignores the impact of kinetics of aggregation, which could lead to potential errors [25]. Some laboratories use n-heptane for precipitation and others use n-pentane. Some precipitate asphaltenes hot and others precipitate them cold. Some use Soxhlet washing and others do not. All of these variables significantly affect the solubility parameters and the way the asphaltenes are partitioned. As a result, it is difficult to compare the asphaltenes generated from one laboratory to another. Consequently, the analytical measurements that follow these incompatible partitioning procedures could lead to confusing or even contradicting results. Therefore, a repeatable method for characterization of asphaltenes is required.

Conventional methods for measuring the asphaltene content of a crude oil follow the definition of asphaltenes that was presented above. These techniques usually involve mixing the crude oil with a titrant such as heptane at a predetermined volume ratio to induce the precipitation of asphaltenes. The volume ratio of 1:30 to 1:50 crude/titrant is commonly used because maximum precipitation occurs around these ratios [26]. Once the asphaltenes precipitate, they are filtered through a filter paper. However, to remove the occluded oil within the asphaltenes' porous network, extra washing with the titrant is needed. Either the filter is manually washed with extra titrant or it is placed inside a reflux extractor. In both cases, as soon as the solid asphaltenes on the filter paper come into contact with the fresh titrant, the assumption of, say, 1:40 v/v ratio becomes completely invalid. This is perhaps one of the greatest misconceptions about the wet chemistry procedure and a major source of operator/laboratory dependency of these methods. In fact, using tools such as reflux/Soxhlet extractors push the volume ratios to the limit of infinite dilution. Since these parameters are not well quantified, they, in turn, result in precision problems. The lack of repeatability and operator dependency of these techniques have forced researchers, as well commercial laboratories, to come up with their own specific versions of these techniques to improve data quality, although at the expense of a lack of an industry-wide standard.

The problems with the complex and labor-intensive wet chemistry procedures have forced researchers to think outside of the box. It turns out that the oil and gas industry is not the only industry grappling with such arduous laboratory procedures. The biomedical laboratories have long been suffering from the limitations of such methods, and their answer to the problem is "lab-on-a-chip" approaches using microfluidic technology. Over the past two decades, microfluidic technology has taken long strides in miniaturizing and automating complex laboratory procedures.

"Micro total analytical systems" or µTAS have been successful in miniaturizing DNA amplification, detection of infectious diseases and chemical warfare agents, flow cytometry, and even cryopreservation. The benefits of these technologies include reduction in samples size, fast turnaround, better process control, portability, automation, operator independency, and low cost [27]. In this chapter, we discuss one of the latest applications of the microfluidic technology in the oil and gas industry. First, we review a brief history of microfluidics, its applications, and basic fabrication techniques. Second, the origin of color in crude oils and principles of spectroscopy will be discussed. Third, the wet chemistry procedure used in this work and the proof-of-concept experiments are explained. Fourth, the principles of fluid mechanics, dispersion, and the operations of the microfluidic system are described. Finally, the chapter is concluded by reviewing the results of precipitation and redissolution tests.

11.2 MICROFLUIDICS

11.2.1 BRIEF OVERVIEW OF MICROFLUIDICS

Microfluidics can be defined as "the science and technology of systems that process or manipulate small amounts of fluids (10^{-9} to 10^{-18} L), using channels with dimensions of tens to hundreds of micrometers" [28]. The field of microfluidics emerged in the 1990s as a natural extension of chromatography and capillary electrophoresis systems. Pioneering work by Andreas Manz and Jed Harrison brought the subject to light [29,30]. Since then, major contributions from thousands of researchers in >300 academic groups and a handful of industrial groups have turned it from an obscure subject to a mainstream field.

Microfluidics is a multidisciplinary subject that draws expertise from both core and applied sciences, including but not limited to fluid mechanics, optics, chemistry, spectroscopy, microelectronics, and microelectromechanical systems (MEMS). However, up until recently, the vast majority of microfluidic applications were in biology and environmental monitoring. Recently, however, there are new applications in the oil and gas and energy sectors.

Biotechnology has been the primary growth area for the field of microfluidics. Its applications were in DNA amplification [31,32], DNA and protein microarrays [33–35], cell biology [36,37], tissue engineering [38,39], and drug discovery [40,41]. Other applications such as fuel cells [42,43] and biofuel cells [44] are expanding as well as environmental monitoring [45,46]. Microfluidic systems enjoy the benefits of small-scale physics. When fluid channels become small, the surface-to-volume ratio increases significantly. This reduction in surface-to-volume ratio allows the surface forces such as interfacial and capillary forces to dominate. Furthermore, in capillaries and microchannels, laminar flows are common. These properties have allowed scientists and engineers to build smaller and more efficient diagnostic systems.

While common definitions of microfluidics do not include the fabrication techniques in the definition, many scientists describe microfluidics only in the context of modern fabrication techniques such as photolithography and chemical etching. In fact, without the modern fabrication methods borrowed from the microelectronics and MEMS industries, modern microfluidics would not exist. The first microfabrication techniques such as wet etching and deep reactive ion etching were borrowed straight from the MEMS and microelectronic industries. In the mid-1990s, the microfluidics community developed fabrication methods, such as soft lithography using polydimethylsiloxane (PDMS), to build devices rapidly and without a cleanroom. These approaches were developed specifically to improve turnaround time and lower the cost in the manufacturing of microfluidic devices. Fabricating micro fluidic devices from PDMS is perhaps one of the most common and cost-efficient techniques for research and development purposes, although it has minimal commercial application because of the lack of tolerance for pressure, temperature, and organic solvents. Conventional macroscopic manufacturing techniques have also been extended for microfabrication successfully. Micromilling, laser ablation, hot embossing, micromolding, and three-dimensional printing techniques are capable of creating microscale features in a variety of materials from stainless steel to plastics and polymers.

This section is not intended to be an exhaustive review of all available microfabrication methods. However, we review the common microfabrication method for glass devices, which is used in this study. Interested readers can see References 47 through 50 for an in-depth review of fabrication methods.

The most common microfabrication technique for planar structures is depicted in Figure 11.1. The method comprises three major steps: deposition, patterning, and etching. In the first step, a thin layer of metal is deposited on a substrate. Then, the desired pattern is transferred onto the deposited layer. Finally, the unwanted parts are removed from the deposited layer. This three-step cycle can be repeated as many times as desired. The microelectronics industry has been making transistor, resistor, and capacitor layers using this technique for decades; we are now using the same principles to make microchannels.

Photolithography is a powerful technique to transfer a pattern on to a substrate. In photolithography, a transparent mask is fabricated using a precise mask writer. The mask is usually a glass substrate with the metal pattern on it. The pattern on the mask is then transferred onto a light-sensitive polymer called photoresist using a short exposure of the photoresist to light through the mask. Figure 11.2 depicts a

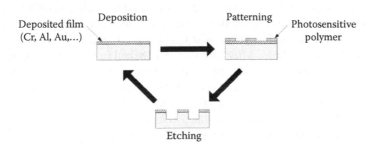

FIGURE 11.1 Basic concept of planar microfabrication involving deposition of a masking layer, patterning channel features, and etching the substrate.

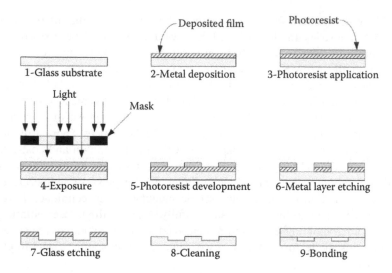

FIGURE 11.2 Simplified steps illustrating the cross-sectional view of microfluidic chip fabrication. The process starts with cleaning of glass substrates and deposition of masking layers. The channel design is transferred to the masking layer by photolithography. Wet chemical etching processes are used to convey the pattern onto the substrate. The final step is bonding two halves forming an enclosed network of channel structures.

schematic diagram for patterning microchannels in glass. The process described in the figure closely resembles the process used to fabricate microfluidic devices discussed in Section 11.4.

The first step in any microfabrication technique is cleaning the substrates to remove surface contamination and particles. Typical particles floating in an office or laboratory environment are at the same scale as the microchannels, which could interfere with the lithography and etching steps. Therefore, most microfabrication processes are conducted in a "cleanroom" environment where the number of particles is reduced using high-efficiency filters. The cleaning is usually conducted by using strong acids such as hot sulfuric acid followed by high-pressure washing with deionized and submicron filtered water. Once the substrates are clean, a thin metal layer, such as chromium, is deposited on the glass substrate. Sputtering and/or evaporation techniques under vacuum are commonly used for well-controlled deposition of metals on substrates. The third step in the figure depicts the application of photoresist on the metal layer. The photoresist is usually a liquid light-sensitive polymer that is spun on the substrate. After spinning, it is baked in an oven to solidify. Once the photoresist is baked, it can be exposed to light through the mask using a mask aligner. The mask is positioned on top of the photoresist as shown in step 4, and the resist layer is exposed to light through the mask. Therefore, the patterns on the mask are transferred to the photoresist. This step is akin to exposing photosensitive paper to light through the negative in conventional film photography. Similar to photography, the photoresist is developed by using a solvent that dissolves the exposed area and leaves the rest of the resist intact, which is shown in step 5. In step 6, the exposed metal layer can be etched using a suitable acid. Similarly, the exposed glass surface can now be etched using an acid such as hydrofluoric acid, as depicted in step 7. Once the channels are etched to the desired depth, the metal layer is completely removed and the etched glass substrate can be cleaned using step 1. By repeating steps 1 through 8, one can add to the complexity of the device. For instance, channels with multiple depths can be fabricated, or electrodes can be fabricated inside the channel.

To close the channels, a top substrate is needed. The top substrate, which may or may not have its own features including drilled access ports to the channels, is thoroughly cleaned and then bonded with the bottom substrate. To make the bond permanent, the glass substrates shown in step 9 can be annealed near the glass transition temperature in a furnace. The fused substrates will typically have multiple chips of the same design. The individual chips are often cut out using a dicing saw. Depending on the budget, such fabrication procedures could be manually intensive or heavily automated.

11.2.2 Applications in the Oil and Gas Industry

Microfluidics is an interdisciplinary field right at the intersection of engineering, physics, chemistry, and biology. However, the life sciences have been the primary beneficiary of this technology. Applications of microfluidic technology in point-of-care sensors and diagnostics have been the main driver behind its explosive growth during the past two decades. Devices have reached large-scale commercialization as demonstrated by the success of products such as the *i-STAT* from Abbott Laboratories, USA [27,51]. However, there are potential applications of microfluidics outside of the life sciences that have yet to be realized. This is evident in the oil and gas industry in particular, where the applications of microfluidics have not gone beyond the traditional micromodels.

Micromodels are perhaps the oldest application of microfluidic systems, well before the word "microfluidics" was even coined. In 1961, two brilliant engineers, Mattax and Kyte, developed the very first microfluidic device in the world [52]. They were interested in displacement of crude oil by water in porous structures for enhanced oil recovery applications. Before their work, such experiments were conducted in sand-packed or glass bead-packed columns. While the packed column resembled the porous structure in the rock, it was difficult to see the interactions between individual water and oil droplets inside the packed bed. Therefore, Mattax and Kyte came up with a planar structure to be able to closely observe such interactions. They took a glass substrate, covered it with wax, and cut through the wax with a stylus. They created a network of 350×350 channels. Then, they etched the channels using hydrofluoric acid. Finally, they bonded another glass substrate on top of it using a fusion bonding technique. Thus, they built the first ever microfluidic device.

Since then, micromodels have been used in studying transport phenomenon in porous structures such as invasion percolation [53] and fluid displacement [54] primarily for enhanced oil recovery applications [55]. Micromodel applications in the real world, however, are limited. They are a powerful observation tool; however, it is difficult to make any quantitative connections between micromodel measurements and the reservoir. Recently, applications of microfluidics in oil characterization and diagnostics have gained momentum. This is a new direction that holds promise. Utilization of microfluidics for interfacial characterization of crude oil films [9,56], asphaltene studies [57–60], determination of solubility and diffusion of gases in crude oil [61–64], measurement of phase behavior [65], and gas–oil ratio [66] are recent examples of this new direction.

11.3 DETECTION: ULTRAVIOLET–VISIBLE SPECTROSCOPY

11.3.1 ORIGIN OF COLOR IN ASPHALTENES AND CRUDE OILS

Microfluidic devices are designed to manipulate small amounts of fluid using microfabricated channels, mixers [67], and valves/pumps [68]. However, the reduced quantity of analyte produced from a microfluidic device requires some consideration with regard to the sensing strategy. Often, optical or electrochemical sensors are preferred as they are sensitive, have quick responses, and are generally scalable. Ultimately, these approaches can be integrated within the chip to yield a small footprint device (think handheld); for example, embedded absorption flow cells [69,70] or integrated CMOS-level components [71]. The simplest optical characterization is coloration, and it becomes a powerful interrogation tool when combined with prior knowledge of the fluid composition and its interaction with light.

The visible colors of typical crude oil samples span a gradient that includes transparent, pale yellow, tan, dark brown, and black [72], as shown in Figure 11.3. To identify the components of crude oil that contribute to its coloration, we must briefly cover light and matter interactions. When a bulk material is exposed to electromagnetic radiation, the incoming energy interacts with the atoms and molecules constituting the bulk material. The nature of the interaction largely depends on (i) the state, composition, and arrangement of the matter, and (ii) the frequency, orientation, and intensity of the incoming light. As the incoming stream of photons/waves bombard the electron cloud of the material, transformation of the incident energy leads to observable phenomena. These encounters can be categorized as either scattering or absorption processes, and both can produce a perceptible color in materials [73]. In fact, ancient and medieval artisans were the first to exploit scattering and absorption effects of nanoparticles to produce stained glass windows and pottery possessing vibrant colors [74].

An absorption event occurs when a molecule "takes up" an incident photon, thereby exciting the molecule to a higher energetic mode. Generally speaking, molecules have three "quantized" energetic modes: electronic (ultraviolet–visible [UV–Vis] wavelengths), vibrational (infrared wavelengths), and rotational (microwave

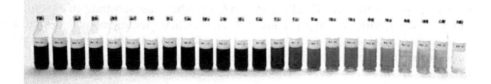

FIGURE 11.3 Several crude oil samples from a single column in an oil reservoir. The vertical profile shows that there is a compositional change detectable by coloration. (Reprinted with permission from Ruiz-Morales, Y. et al., Electronic absorption edge of crude oils and asphaltenes analyzed by molecular orbital calculations with optical spectroscopy, *Energy Fuels, 21* (2), 944–952. Copyright 2007 American Chemical Society.)

wavelengths) [75]. The incoming photon energy must match the energy quanta required to transition the molecule from a ground state to an excited state. This is known as a characteristic frequency or absorption line, and molecules often have several that give rise to an absorption spectrum. Molecular absorption processes are governed by the conservation of energy, and the imparted energy must then be dissipated. This can be achieved through various pathways, some of which are graphically shown using the familiar Jablonski diagram in Figure 11.4.

Crude oils are commonly separated into four bulk fractions: saturates, aromatics, resins, and asphaltenes (SARA). The saturate fraction is nonpolar and is composed of normal alkanes, branched alkanes, and cycloalkanes. Saturates tend to absorb strongly in the deep UV, with high-frequency electronic transitions, as their electrons are tightly bound and require more incident energy to be excited (wavelengths <200 nm). As alkanes increase in carbon number, there is a notable red shift. For instance, the absorption edge of pentane is 173 nm, whereas the absorption edge of decane is 176 nm, a red-shift of 3 nm [76]. This bathochromic shift is presumably due to a smaller energetic gap between the highest occupied molecular orbital (HOMO) and lowest unoccupied molecular orbital (LUMO). Costner et al. conclude that the LUMO energy level is relatively independent of carbon number and that it is the HOMO level that increases with increasing carbon number.

The aromatic hydrocarbon fraction is more polarizable and contains molecules that have alternating double and single bonds between carbon atoms, producing conjugated electron orbital systems. Aromatic hydrocarbons form structures of one or more rings, where multiring structures are often referred to as polycyclic aromatic hydrocarbons. Similar to saturates, a red shift is noted as complexity increases and more rings are added to the molecule; the electron field becomes more delocalized from overlapping π-bonds.

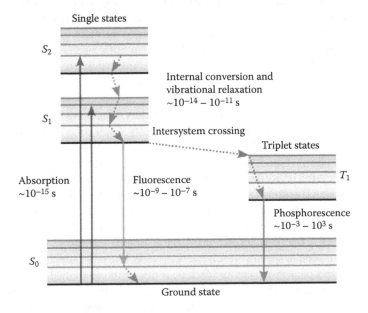

FIGURE 11.4 Jablonski energy diagram.

The presence of substituents on the ring structures can also cause bathochromic or hypsochromic shifts of tens to hundreds of nanometers. The bathochromic shift of aromatics is much more dramatic than saturates; for instance, the characteristic absorbance peak shifts from 255 nm for benzene (one ring), to 286 nm for naphthalene (two rings), to 375 nm for anthracene (three rings), and to 477 nm for tetracene (four rings) [77–79]. Aromatics absorb strongly in the UV–Vis and tend to have a yellow/orange color.

The last two SARA fractions, resins and asphaltenes, are alkylated aromatic polycyclic clusters with incorporated alkyl side chains, heteroatoms (N, S, O), and trace metals (Ni, V, Fe) [1]. They are the most polar fractions of crude oil. The difference between these two fractions is in their solubility profile. Resins are defined as being soluble in n-alkanes such as pentane or heptane. Conversely, asphaltenes are arbitrarily defined as being insoluble in light n-alkanes but soluble in toluene or dichloromethane [80]. For the purpose of understanding coloration in oil, we consider the molecules that comprise resin–asphaltene fractions as a molecular continuum and employ the Yen–Mullins molecular model for guidance. In this model, fused aromatic ring systems are stacked together, forming nanoaggregates. These aggregates then cluster, forming colloids that are typically around 5 nm in size [2]; recent research indicates that asphaltene nanostructures are polydispersed with sizes as large as 200 nm [81]. The color of these fractions would derive largely from the absorption of the underlying asphaltene molecules with modification from light scattering of the dispersed nanoaggregates. Derakhshesh et al. [82] demonstrated that asphaltene samples in toluene could scatter light significantly and follow a classical λ^{-4} dependence, which indicates a Rayleigh scattering effect. They suggested that scattering was the primary "absorbance" mechanism for wavelengths >550 nm over a wide range of concentrations and from varying geographical origins. Conversely, Ruiz-Morales et al. [72] showed spectral data for many crude oil and asphaltene samples that do not have optical scattering and yet have notable coloration beyond 600 nm. They suggested that asphaltene aggregates contribute to coloration primarily through absorption mechanisms. It is likely that both mechanisms contribute to oil coloration; however, the extent of each effect is unclear at present.

A crude oil sample may have hundreds to thousands of distinct asphaltene molecules with a unique distribution. These alkylated aromatic ensembles may include 20 or more fused rings [83]; however, recent evidence indicates that there is a convergence to an average structure centered at roughly seven rings and having a width ranging from 2 to 20 nm [2,72]. Molecular orbital calculations show that seven-ring structures can absorb broadly between 300 and 700 nm depending on the ring configuration [84]. Also important is the ratio of carbon atoms in isolated double bonds to those in aromatic sextets [72]. As with saturated and aromatic hydrocarbons, resins and asphaltenes show a red shift as the number of fused aromatic rings increases, and also when the ratio of isolated double bonds to those in sextets increases.

When all SARA fractions are combined, the coloration that results will be the superposition of the molecular spectra in each fraction, neglecting molecular interaction effects. The UV–Vis absorption spectrum for black oil is often exponentially decaying from UV toward Vis wavelengths. This characteristic decay can be described as an Urbach edge, which, when plotted in a different way, shows a positively linear relationship between photon energy and the log of absorption. The Urbach edge for multiple crude oil types is summarized in Figure 11.5 [72]. Interestingly, both crude oils and asphaltenes exhibit similar slopes, or "Urbach" behavior. For the oils shown in Figure 11.5, Ruiz-Morales et al. propose that the characteristic decay width or slope arises from the population distribution of molecules. Essentially, larger cyclic species exist in exponentially diminishing concentrations and have broad absorption bandwidths. Consequently, when measuring UV–Vis coloration along the linear part of the Urbach tail or absorptive edge (approximately 600 nm), we can estimate the concentration of asphaltenes in the crude oil sample.

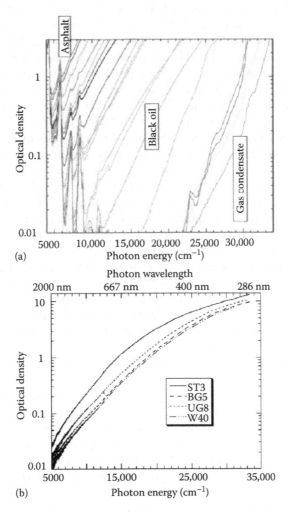

FIGURE 11.5 The log of optical density (OD) versus photon energy, highlighting that for many crude oils (a) and asphaltenes (b), the electronic absorption edge is characterized by a similar slope. Ruiz-Morales et al. describe the origins of this "Urbach" behavior as arising from the exponentially decreasing population of larger cyclic species. (Reprinted with permission from Ruiz-Morales, Y. et al., Electronic absorption edge of crude oils and asphaltenes analyzed by molecular orbital calculations with optical spectroscopy, *Energy Fuels, 21* (2), 944–952. Copyright 2007 American Chemical Society.)

11.3.2 Wet Chemistry Procedure

Crude oil samples, especially near the critical points, are known to undergo gravitational settling within reservoirs. Settling of asphaltenes also occurs in sample containers in the laboratory. If a sample contains suspended asphaltene aggregates or other solid phases such as wax, gravitational settling could affect subsampling of the crude oil container. Our observations show that compared with gravimetric techniques, optical measurements are more prone to errors and repeatability problems due to such a small concentration gradient. Therefore, for a successful optical-based measurement, a consistent and reliable subsampling procedure is of prime importance. Kharrat et al. [57] showed that conditioning the sample at 60°C for 16 h would result in a homogeneous sample. The same procedure is also followed here.

To determine the asphaltene content using the gravimetric technique, a modified version of ASTM D6560 [26] method was used. For each oil sample, a 1-g aliquot was mixed with 40 mL n-heptane. The mixture was left for 2 h for the asphaltene molecules to aggregate. The mixture was then filtered using a 200-nm Teflon filter. The filter was folded and placed in a Soxhlet extractor. The Soxhlet was used to wash the asphaltenes on the filter with hot heptane and remove all non-asphaltene material such as resins, aromatics, and wax. The washing continued until heptane was clear. The filtrate material was redissolved in 100–150 mL of high-performance liquid chromatography–grade dichloromethane through the filter. The dichloromethane was later removed using a rotary evaporator until the solution was concentrated to approximately 10 mL. The concentrated solution was transferred to a small vial and placed on a hot plate under nitrogen for the remaining solvent to evaporate. Once the solvent evaporated, the vial was weighed every 15 min until the mass of the vial reached a constant value. The concentration of asphaltenes was calculated on the basis of the mass of asphaltenes in the vial and the original mass of the sample aliquot.

Unfortunately, it is nearly impossible to determine the uncertainties associated with the wet chemistry technique described above. ASTM D6560 suggests a repeatability of <10% and a reproducibility of <20%. The standard also suggests that 1 sample of 20 will fall outside of these uncertainty limits. However, the suggested uncertainty is based on an interlaboratory study on only four samples, which is a limited data set [26]. A number of factors make the uncertainty of asphaltene content measurements strongly dependent on the sample and its asphaltene content. A repeatability of <10% or reproducibility of <20% is perhaps reasonable for highly asphaltenic fluids. However, for fluids with low asphaltene contents, say <1%, the uncertainty is expected to increase. Other factors such as concentration and size distribution of the inorganic phase (clay) present in any sample, and the stability of the asphaltene fraction make the wet chemistry method a highly sample-dependent measurement. Furthermore, environmental parameters such as temperature and humidity of the laboratory, Soxhlet wash duration, and operator skills play an important role in the repeatability and reproducibility of the measurement.

Since the repeatability of the wet chemistry method highly depends on the composition and asphaltene content of the sample, the best method to establish the uncertainty is to repeat every measurement multiple times. Multiple measurements for a sample, however, is prohibitively expensive and time consuming. For a series of samples, Kharrat et al. [57] reported 10%, 20%, and 50% standard deviation in the estimates of asphaltene content, for fluids with asphaltene content of <1%, between 1% and 3%, and >3%, respectively. We will be using these numbers as a guide for the magnitude of wet chemistry errors in this study.

11.3.3 Proof of Concept Using Centrifugation

Asphaltenes are known to follow the Beer–Lambert law in organic solvents [7]. The Beer–Lambert law states that there is a linear correlation between optical absorbance and concentration:

$$\text{Absorbance} = A = -\log_{10}\left(\frac{I}{I_0}\right) = \varepsilon.c.l$$

where I, I_0, ε, c, and l represent the intensity of the transmitted light, intensity of the incident light, molar absorptivity, concentration, and optical path length, respectively. The Beer–Lambert law applies at the limit of infinite dilutions where it is assumed molecules do not interact with each other. However, asphaltenes are known to self-associate and form aggregates, which may cause nonlinearities in absorbance and potential error in measurement at a high concentration of asphaltenes. Therefore, care must be taken to use the Beer–Lambert law limited to low asphaltene-content fluids. Our intention is to investigate heavy oils and bitumen at a later date and use this method for only black crude oils at present.

Figure 11.6 shows a correlation between the concentration of asphaltenes and their absorbance in toluene. The vertical axis shows the difference in absorbance at 600 and 800 nm, and the horizontal axis represents the concentration of asphaltenes. In this chapter, all absorbances are presented as a difference between the optical

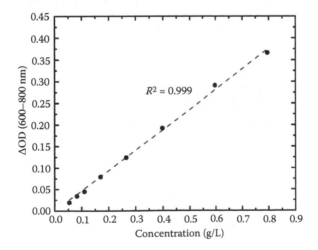

FIGURE 11.6 Linear correlation of absorbance versus asphaltene concentration.

densities at 600 and 800 nm. By calculating the difference at the two wavelengths, we minimize the errors due to baseline shift. The 600-nm wavelength was chosen because of the overlapping linear Urbach behavior of many black oils and asphalts at that wavelength. Additionally, the absorbance at 600 nm fit within the dynamic range of the flow cell and spectrometer for the entire sample set. At 800 nm, the spectrometer had a relatively good signal-to-noise ratio enabling baseline correction. Depending on the sample set, spectrometer, flow cell, and light source, other wavelengths can be used. Other methods, such as calculating the derivatives or integrals of the spectra, can also be used.

The strong linear correlation in the figure indicates that the Beer–Lambert law can be applied to asphaltene solutions. Conversely, the Beer–Lambert law can be exploited to determine the concentration of asphaltenes in a crude oil.

In a real crude oil, however, there are other species that absorb in the same wavelengths. For instance, the resin and aromatic fractions absorb in the UV–Vis range, which interfere with the absorbance of asphaltenes. Therefore, a method of separating the spectrum of asphaltene from the rest of the crude oil is needed.

Here, we use the principle of superposition and split the spectrum of crude oil into that of asphaltenes and maltenes (deasphalted fraction), as depicted in Figure 11.7. The superposition principle is a valid assumption here since the precipitation of asphaltenes is a physical separation based on solubility parameter, not a chemical one. We also assume that the impact of molecular interactions between asphaltenes and other species in the crude oil is negligible, which is an inherent assumption when we use the Beer–Lambert law.

Figure 11.8 shows typical UV–Vis spectra for a crude oil and its associated asphaltenes and maltenes. For this sample, the asphaltenes were precipitated using the wet chemistry method explained in Section 11.3.2. Then, the asphaltenes were diluted in 40 mL toluene. The crude sample (1 g) was also diluted in 40 mL toluene. The maltenes, however, were diluted in 40 mL heptane. The graph shows that the superposition principle is a reasonable assumption for this crude oil.

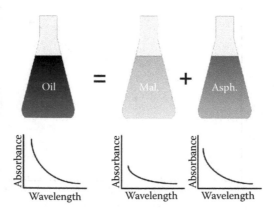

FIGURE 11.7 Principle of optical absorption superposition. The spectrum of oil is the summation of the asphaltene spectrum and the maltene spectrum.

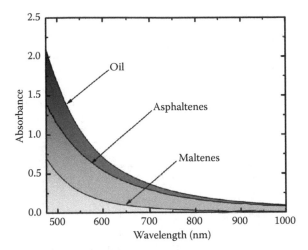

FIGURE 11.8 Measured oil, maltene, and asphaltene spectra diluted with a solvent at 1:40 (g/mL).

The method explained above, however, is not practical. This method requires precipitation and separation, redissolution in a solvent, and finally, spectroscopy of asphaltenes. In other words, one would be required to go through all of the wet chemistry steps first before the spectroscopic measurement can be performed, which would in turn bring all the errors associated with the wet chemistry technique directly into the spectroscopic measurements. Therefore, we use the superposition principle and measure the spectrum of the asphaltenes by subtracting the spectrum of maltenes from that of the crude oil.

By subtracting the spectrum of maltenes from the crude oil, we can determine the spectrum of asphaltenes indirectly, without any mass measurement or redissolution in a solvent. Figure 11.9 shows a schematic diagram for such a system. As depicted in the figure, there are two steps in the process: measuring spectrum of crude oil and spectrum of maltenes.

In the first step, the spectrum of the crude oil is measured and recorded. Then, the crude is mixed with heptane at 1:40 v/v, which induces the precipitation of asphaltenes. Once the aggregates are formed, they can be filtered using a filter paper or a centrifuge. The UV–Vis spectrum of the permeate (or the supernatant in case of centrifugation) is then measured. By subtracting the spectrum of permeates from that of the crude oil, the spectrum of asphaltenes can be deduced. Kharrat et al. showed that the absorbance of asphaltenes correlates well with their mass concentration measured using the wet chemistry method. A similar approach was used by Bouquet and Hamon in 1985 [85].

Figure 11.10 shows the results from Kharrat et al., where they used a 2-mm pathlength cuvette for their measurements. The horizontal axis shows the spectrum of asphaltenes, which is calculated on the basis of the difference in the absorbance of oil and maltenes. The vertical axis shows the asphaltene content using the gravimetric technique. Kharrat et al. also incubated their samples for 2 h after mixing with

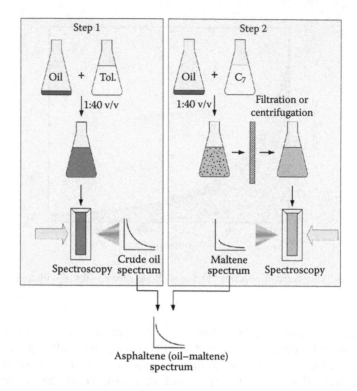

FIGURE 11.9　Schematic diagram for measuring the optical spectrum of asphaltenes by the subtraction method.

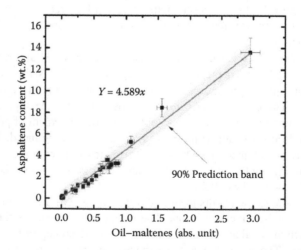

FIGURE 11.10　Correlation between optical absorbance and wet chemistry asphaltene content.

heptane for the asphaltene aggregates to grow, followed by centrifugation for separation. They took an aliquot from the top of the centrifuge tube for spectroscopy without agitating the mixture. For the spectroscopy of the crude oil, they mixed the crude sample with toluene at a 1:40 v/v ratio. The figure shows a good correlation between the absorbance of asphaltenes and their concentration, albeit the data set is sparse at high concentrations of asphaltenes. With the optical sensing strategy of asphaltene concentration now enabled, we direct our attention toward the background of fluid flow in microchannels.

11.4 FLUID FLOW IN MICROCHANNELS

11.4.1 Fluid Mechanics Fundamentals

Fluid flow in capillaries and microchannels is a rich and exciting topic. It includes single-phase compressible/incompressible flows, multiphase flows (gas–liquid, solid–liquid, and liquid–liquid), heat and mass transfer (evaporation, boiling, condensation), electrohydrodynamics (electro-osmosis, electrophoresis), and magnetophoresis, to name a few. In this section, however, we briefly focus on the fundamentals of the fluidic system discussed in the chapter, which include incompressible laminar flow, mixing, and dispersion in microchannels.

In the microfluidic system that will be described in detail in Section 11.5, we use positive displacement pumps (syringe pumps) to inject a hydrocarbon at a predetermined and constant flow rate, which allow us to simplify the problem considerably. Therefore, we can safely use the Poiseuille flow model. At first, we only consider the Hagen–Poiseuille solution for circular capillaries. Then, we will extend it to microchannels with noncircular cross sections. Assuming that the flow is fully developed, axisymmetric, and that all flow is in the z-direction (streamwise direction), the Navier–Stokes equation can be written as

$$\rho \frac{\partial \vec{u}}{\partial t} + \rho \vec{u}.\nabla \vec{u} = -\nabla p + \mu \nabla^2 \vec{u} \qquad (11.1)$$

where ρ, $\vec{u}$, p, and μ represent density, velocity vector, pressure, and viscosity, respectively. The above equation can further be simplified to

$$\nabla p = \mu \nabla^2 \vec{u} \qquad (11.2)$$

or

$$\frac{\partial p}{\partial z} = \mu \frac{1}{r} \frac{\partial}{\partial r} \left(r \frac{\partial u_z}{\partial r} \right) \qquad (11.3)$$

where r is the radial coordinate and u_z is the streamwise velocity component [86]. Assuming no-slip boundary conditions and finite flow velocity at the center of the capillary, we obtain the famous parabolic flow velocity profile for circular tubes:

$$u_z = -\frac{1}{4\mu}\frac{\partial p}{\partial z}(R^2 - r^2)$$ (11.4)

where R is the capillary radius. Using the above equation, one can calculate the flow rate and the average flow velocity:

$$Q = -\frac{\pi R^4}{8\mu}\frac{\partial p}{\partial z}, \quad \bar{u}_z = \frac{1}{2}u_{z_{r=0}} = -\frac{\partial p}{\partial z}\frac{R^2}{8\mu}$$ (11.5)

For a noncircular channel, the pressure drop equation can be generalized to

$$\frac{dp}{dz} = -f\frac{\rho\bar{u}_z^2}{2D_h}$$ (11.6)

where f is the Darcy friction factor and D_h is the hydraulic diameter. For a circular channel, the friction factor is $f = 64/Re$, where $Re = D_h\rho\bar{u}_z/\mu$ is the so-called Reynolds number, which is the ratio of inertial forces to viscous forces. For low Reynolds number flows, the product of the friction factor (f) and the Reynolds number (Re) is called the Poiseuille number (Po), which is a function of channel geometry (Po = f.Re). Therefore, the Poiseuille number for circular channels is $Po = 64$. For microchannels with square cross section, the Poiseuille number is $Po = 56.92$. For rectangular channels, the following correlation can be used [87–91]:

$$Po = f.Re = 96(1 - 1.3553\lambda + 1.9467\lambda^2 - 1.7012\lambda^3 + 0.9564\lambda^4 - 0.2537\lambda^5)$$ (11.7)

where λ is the aspect ratio of the channel cross section. The nature of the cross-section geometry of a microchannel depends on the material and the manufacturing techniques available to create that channel. Microchannels etched in silicon are close to square or rectangular at shallow depth and trapezoidal when the channel depth is large. For microchannels fabricated in glass substrates, owing to isotropic etching, the channel usually has a semicircular shape. The Poiseuille number for more complex geometries can be found in References 88 and 89.

The equations we have derived thus far are valid for laminar flow regimes, which is the primary focus of this section. The transition from laminar flow to turbulent flow is determined by the Reynolds number. In the 1890s, Osborne Reynolds showed that for pipe flows, the transition from laminar flow occurs at Re <2300. In circular capillaries, recent experiments show a transition region from Re = 1800 to 2000, which largely support the macroscopic observations. For noncircular microchannels, however, the results are inconclusive and further measurements are required [92].

In the microfluidic system that will be discussed here, the fluid travels through polytetrafluoroethylene (PTFE) tubes, microchannels etched in glass, and micro-/

minichannels in stainless steel. The maximum Reynolds number is <30, which can be safely considered laminar, and therefore, the analysis presented above is valid.

11.4.2 MIXING IN MICROCHANNELS

As we described in Section 11.4.1, in a wide variety of applications, our case included, the Reynolds number in the channel is small enough so that the convective terms of the Navier–Stokes equation are small. In such circumstances, in the absence of eddies, swirls, and cross currents perpendicular to the direction of flow, diffusion is the dominant mechanism for mixing. For a given distance ℓ, the time required for diffusion to take place is $\tau = \ell^2/D$, in which D is the diffusion coefficient. In the absence of convective transport, the diffusion time could be prohibitively long. For instance, the time required for a typical hydrocarbon to diffuse along a 1-mm channel in hydrostatic condition is in minutes. Therefore, for efficient mixing, convective transport will have to be introduced into the fluidic system.

One of the most effective ways to improve mixing in microfluidic systems is by shortening the diffusion length using the flow-folding technique. The method is borrowed from bakers who mix a viscous fluid, the dough, by repeated folding and stretching. Figure 11.11 shows the baker's transformation. By repeated folding and stretching, the diffusion length is reduced exponentially. Therefore, the diffusion distance is reduced substantially [86,93].

Micromixers that are based on such principles are chaotic [94,95]. Chaotic systems, unlike random systems, are deterministic systems. In other words, we can model the chaotic systems and predict their behavior to some extent in the future. However, they are extremely sensitive to initial conditions, and owing to their highly nonlinear nature and exponential growth, minute fluctuations in the initial conditions will have a dramatic impact on the output.

Figure 11.12 shows the type of micromixer used in this study (Dolomite, UK). The micromixer is etched in glass because of its compatibility with hydrocarbons. The mixing principle again is based on split-and-fold to reduce the diffusion length and increase the interfacial area between the two fluids. In the figure, at the Y-junction, the crude oil is injected through the middle channel. The titrant, which is n-heptane, is injected from the two side channels. Asphaltene particles precipitate right at the

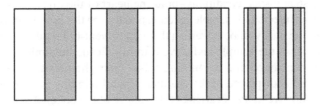

FIGURE 11.11 Baker's transformation is a process that enhances mixing of two parallel laminar flows by splitting and recombining the flows in an overlapping manner. Small interstream path lengths lower the molecular diffusion time.

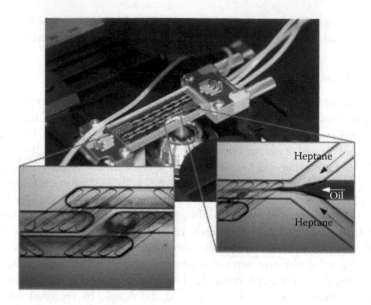

FIGURE 11.12 Example of a micromixer in action, mixing crude oil with n-heptane. The Y-junction introduces the three streams, while several stages of mixers merge the fluids. The small channel cross sections in the figure are 50 μm × 125 μm.

interface of oil–heptane as soon as the two fluids come into contact. Once formed, the particles at the interface could delay the mixing and precipitation. To enhance mixing, at each stage of the mixer, the flow is split into eight streams using the narrow diagonal channels before they are merged into a main channel again. This process can be repeated as many times as desired.

11.4.3 DISPERSION

The microfluidic device that measures asphaltene absorbance will be based on a customized flow injection system that injects a slug of oil sample into a carrier stream of solvent. As the carrier solvent moves the plug, smearing or broadening of the injected square plug will occur. Therefore, the measurement should be timed to take a reading at a steady-state condition—or at the heart of the mixture plug, where the concentration of injected oil sample is representative of the initial sample.

The oil sample plug will be travelling through solvent-filled tubes and microchannels that will cause the sample and solvent to mix at the beginning and end of the plug. The mechanism behind the mixing arises from the pressure-driven flow system. As the fluid is displaced in the channel, a parabolic flow profile develops with the maximum flow velocity occurring at the center of the channel and the minimum flow at the boundaries or channel walls. Radial diffusion of molecules across the cross-sectional varying flow velocities leads to enhanced mixing at the boundary, otherwise known as shear-enhanced diffusion. For this type of dispersion to occur, the characteristic radial diffusion time must be small compared with the transit time of the plug in the channel. This is known as Taylor–Aris dispersion, and it has a significant effect on plug broadening in small channels. It is mathematically expressed with Equation 11.8, which describes the concentration profile both spatially and temporally [96,97]:

$$\frac{\partial c}{\partial t} + \bar{u}_z \frac{\partial c}{\partial z} = D_{\text{eff}} \frac{\partial^2 c}{\partial z^2} \tag{11.8}$$

where c is the molecular concentration (mol/m³), t is time (s), z is the position (m), $\bar{u}_z$ is the mean flow velocity (m/s), and D_{eff} is the "enhanced diffusion factor" or the Taylor–Aris dispersion coefficient, which can be calculated for capillary channels as follows:

$$D_{\text{eff}} = D_{\text{m}} + \frac{d^2 \bar{u}_z^2}{192 D_{\text{m}}} \tag{11.9}$$

where d is the pipe diameter (m) and D_{m} is the molecular diffusion coefficient. For arbitrary channel cross sections, please see Reference 98. The first term in Equation 11.9 is the contribution from ordinary axial diffusion and the second term is the enhanced dispersion from the flow profile. In microfluidic applications, the axial diffusion is typically small compared with the "enhanced diffusion factor" of the convective flow profile. Therefore, using Equation 11.9, the Taylor–Aris dispersion will be less pronounced with smaller diameter channels, slower flow velocities, and small–rapidly diffusing molecules (increased D_{m}).

In the case of a flow injection analysis system, we are interested in predicting the shape of the output signal that arises from a boxcar injection. The normalized signal intensity can be determined from a solution to the Taylor dispersion equation above [58], calculated as

$$\left.\begin{array}{ll} c = C_0 & \text{for } -\dfrac{L_d}{2} \le z \le \dfrac{L_p}{2} \\[2ex] c = 0 & \text{otherwise} \end{array}\right\} \tag{11.10}$$

$$I(t) = \frac{1}{2}\left[\text{erf}\left(\frac{L_d - \bar{u}_z t + L_p/2}{2\sqrt{D_{\text{eff}} t}}\right) - \text{erf}\left(\frac{L_d - \bar{u}_z t - L_p/2}{2\sqrt{D_{\text{eff}} t}}\right)\right] \tag{11.11}$$

where L_d is the length to the detector (m) and L_p is the total plug length (m). The cartoon in Figure 11.13 shows the process of Taylor dispersion occurring on a square plug at various times during its transit through the channel. Above each channel is a snapshot of the concentration profile versus the axial spatial dimension from the center of the plug. This moving frame of reference pictorially shows that Taylor dispersion is mathematically a symmetrical process that occurs equally on the front and

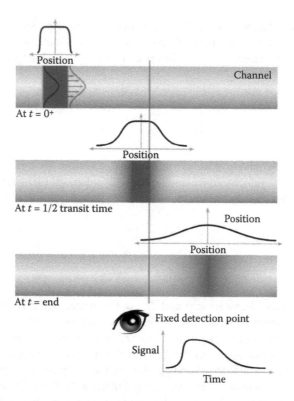

FIGURE 11.13 Square injection plug that will broaden from Taylor–Aris dispersion. The plug shape is shown at various times during its transit through the channel. For a fixed frame of reference, like a detector, slight asymmetry is to be expected.

back edges of the plug. Also shown is a fixed frame of reference, which is often the case when a detector is placed downstream of the injection site. As the plug passes the fixed detection zone, it is simultaneously broadening in time and a slight asymmetry is observed in the output signal. Further experimental factors such as the fluid properties (viscosity mismatch, suspended particles, and intermolecular interactions) and the channel properties (surface roughness, wall interactions, and wetting properties) contribute additional asymmetry.

The various parameters of the asphaltene system were selected to reduce dispersion and make the heart of the plug larger. This included optimizing the flow rate, tubing and channel dimensions, plug length, detection point, and analysis time. The design process starts by understanding the injected sample and in particular the diffusion coefficient of the molecules comprising the sample. Crude oil is a complex mixture of many molecules, and a representative molecule along with its corresponding diffusion coefficient should be selected for a first-order approximation. Since Taylor dispersion is more significant for big–slowly diffusing species, asphaltenes were chosen as they are the largest molecular component

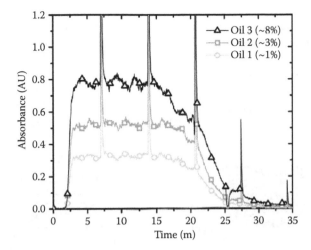

FIGURE 11.14 Square oil plug injection and observed dispersion for three oils with varying asphaltene content. The impulse spikes are the result of the solvent syringe pump refilling. (Reprinted with permission from Schneider, M. H. et al., Measurement of asphaltenes using optical spectroscopy on a microfluidic platform, *Anal. Chem.*, *85* (10), 5153–5160. Copyright 2013 American Chemical Society.)

of the oil. The measured diffusion coefficient for asphaltenes is approximately $D_m = 5 \times 10^{-6}$ cm^2/s [99], meaning that they will diffuse across the channel cross section in seconds. The resident or transit time of the plug is several minutes; therefore, we are within the Taylor–Aris dispersion limits. By adjusting the system parameters, minimal dispersion was achieved. Figure 11.14 shows the dispersion as measured for three samples in the final asphaltene system. The plug leading edge shows a small amount of dispersion, whereas the trailing edge dispersion is more pronounced. In total, only a minor fraction of sample is lost on the plug edges and the majority of the plug shows behavior representative of the injected oil concentration.

11.5 MICROFLUIDIC SYSTEM

A custom microfluidic chip was manufactured in glass, as described in our previous work [58]. A three way Y-junction was used for introducing the sample and the solvent with several stages of micromixers to implement laminar flow folding of the solution. After sufficient mixing, the solution was passed into a long serpentine channel that created a time delay, to ensure that precipitation occurred. The flow was then delivered to an open-face filtration channel, which interfaced with a 0.2-μm PTFE filter membrane for asphaltene separation. The microfluidic chip was enclosed in a cartridge that incorporated external fluidic connections and temperature control, as shown in Figure 11.15. The total dead volume of the block assembly was approximately 150 μL.

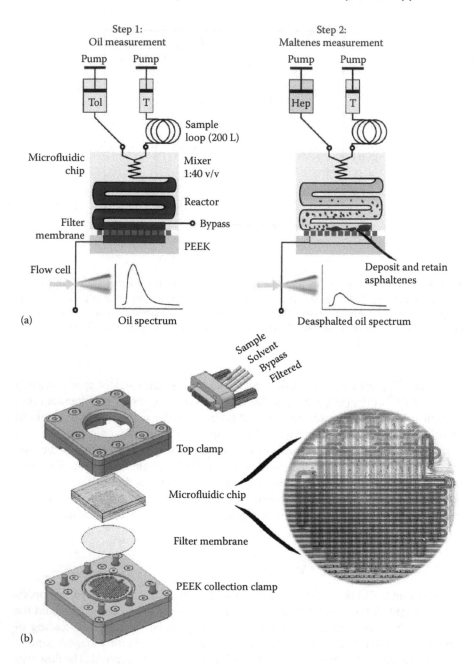

FIGURE 11.15 (a) Overview of the chip-based asphaltene absorbance measurement. The process indirectly measures asphaltenes absorbance by taking the difference of the oil spectrum and the deasphalted oil spectrum. (b) Exploded view of the chip cartridge, holder, and filter. Image on the right is a photograph of asphaltene aggregates in the microchannel during the separation process. (Copyright 2013, Society of Petroleum Engineers Inc. Reproduced with permission of SPE. Further reproduction prohibited without permission.)

The microfluidic chip was interfaced with several other components to produce a completely automated petroleum asphaltene measurement system. To avoid microchannel blockages, both the sample and solvents were filtered using in-line cartridges. Displacement and control of fluids was achieved with three syringe pumps and two rotary valves. Pressure sensors were placed strategically to ensure correct operation and also to detect transmembrane clogging issues. The optical absorption of output fluid was measured with a 2.5-mm path-length flow cell, white-light source, and a UV–Vis spectrophotometer. All subsystems were integrated and automated via a personal computer running LabVIEW 2011 (National Instruments, USA).

The fluidic aspects of the system were reviewed in a previous work [58], namely tuning of flow rates, flocculation kinetics, membrane degradation, and channel clogging. Every experimental run on the system produced a summary data log, as shown in Figure 11.16. Key parameters including microfluidic chip temperature, transmembrane pressure, and output optical absorbance are monitored during the entire

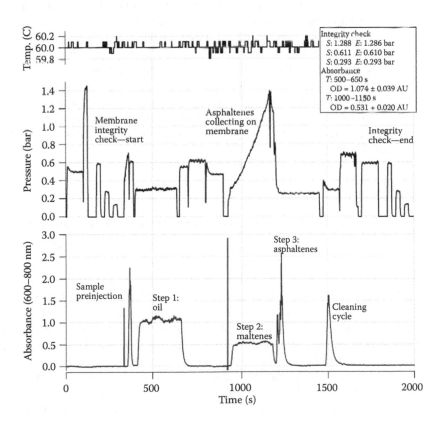

FIGURE 11.16 Example of a summary data log for a crude oil sample run on the microfluidic asphaltene system. These log files are used to troubleshoot and provide quality control on the experimental data. (Copyright 2013, Society of Petroleum Engineers Inc. Reproduced with permission of SPE. Further reproduction prohibited without permission.)

protocol. In this example, the crude oil had an asphaltene content of 3.3 wt.% and a corresponding asphaltene optical absorbance of 0.543 absorbance units (au).

The summary data logs are also used to validate correct system operation. In the protocol established, the first two pressure pulses indicate that the system was successfully primed with toluene. If there is a leak or bubbles, the pressure values would oscillate and be under the expected threshold. Other details, including membrane integrity, can be monitored. For this test, a sequence of three toluene pulses is sent into the system and the average pressure for each of the pulses is recorded as the initial reference. At the end of the run, three more toluene pulses are sent and their induced pressures are recorded. If the pressures match the start of the run, then no tearing or fatal clogging of the membrane had occurred, which indicates the optical data can be more confidently relied on.

Figure 11.16 also shows that the sample properly entered the system and went through the correct steps for analysis. After priming the system with solvent, a sample preinjection is performed. This moves the sample plug from the loading loop to the Y-junction on the microfluidic chip, noted by the optical absorbance spikes before the oil measurement. The sample is then mixed with toluene and the oil absorbance measurement is conducted on a stable plateau, avoiding Taylor-dispersed regions. After more cleaning and priming pulses, the oil is deasphalted by mixing with *n*-heptane. In this step, the absorbance of the filtered maltenes is measured on the stable optical plateau and we note an increasing pressure, indicating that the membrane is collecting asphaltene aggregates. Finally, the deposited asphaltenes are flushed out with toluene, indicated by the sharp increase in optical absorbance and reduction in pressure. These experimental quality checks of the summary data log ensures properly reported values.

11.6 RESULTS

11.6.1 CORRELATION BETWEEN ASPHALTENE CONTENT AND OPTICAL ABSORBANCE

The sample set used to evaluate the efficacy of our device contained 52 crude oil samples from around the world. The samples were all black oils; no heavy oil was included. The crude samples covered a wide range of geographic locations. Figure 11.17 shows the plot of optical absorbance of asphaltenes (oil–maltenes) using the microfluidic device versus the asphaltene content measurements using the gravimetric method explained in the Section 11.3.2 [100]. The vertical error bars show the errors associated with the wet chemistry measurements from Kharrat et al. As discussed in Section 11.3.2, the gravimetric errors were determined to be 10%, 20%, and 50% for samples with asphaltene content >3%, between 1% and 3%, and <1%, respectively. The horizontal error bars show the standard deviation of the average for the heart of the plug. The average is recorded in the center 50% of the plug, and there are inherent noise sources that lead to minor variations in the plateau region. The primary source of these fluctuations is from the fluid flow pulsations produced by the discrete stepping of the syringe pumps. The mismatched pulsations lead to minor and localized variations in dilution when the heptane and oil are combined

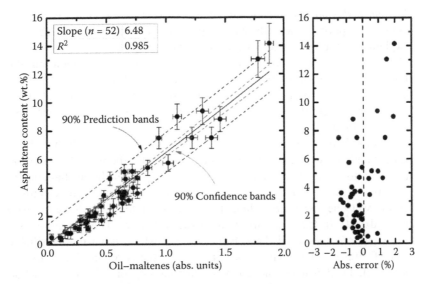

FIGURE 11.17 Correlation of indirect optical absorbance versus asphaltene mass content along with absolute errors from the correlation slope. The x-axis is the difference in oil and maltene absorbances, where absorbances are calculated as the absorbance at 600 nm minus the absorbance at 800 nm. (Copyright 2013, Society of Petroleum Engineers Inc. Reproduced with permission of SPE. Further reproduction prohibited without permission.)

at 40:1 in the chaotic micromixers. Additional noise from the optical system may include mechanical vibrations, thermal noise, light source drifts, and detector-based noise (readout, shot, dark, and fixed pattern).

The plot shows a strong linear correlation between the gravimetric asphaltene content and the absorbance of the asphaltenes. The slope of the linear model fit to the data is 6.48 with an R^2 value of 0.985. The dashed lines show the 90% prediction bands. Schneider et al. [58] reported a slope of 5.46 for similar measurements. However, the difference in slopes is primarily attributed to the broader data set reported in Sieben et al. [100]. The former measurements had only 32 samples, all of them with <8 wt.% asphaltene content, whereas the latter included >50 measurements up to 14 wt.% asphaltene content. The calibration correlation for the microfluidic apparatus can be written as follows:

$$M_{Asp} = 6.48\,A$$

where M_{Asp} shows the concentration of asphaltenes (wt.%) and A represents the optical absorbance for a 2.5-mm path-length flow cell. The figure also shows the relative error for the microfluidic measurements. For the low-asphaltene-content fluids, the error is rather large, whereas for the high-asphaltene-content samples, the relative error is small. The source of the error can be attributed to the following:

- *Wet chemistry measurements.* The repeatability of the wet chemistry measurement is poor, particularly at low asphaltene concentrations, which could explain the relatively large deviations at that region.
- *Variation in extinction coefficients.* Asphaltenes are defined on the basis of their solubility properties, rather than their molecular structure. Therefore, a large and diverse spectrum of molecules with different molar absorptivity can fall under their umbrella, which makes their characterization difficult.
- *Procedural differences between the two methods.* While both the microfluidic and the wet chemistry techniques separate the asphaltenes using precipitation, the microfluidic technique precipitates the asphaltenes at a 1:40 v/v ratio, which is accurately controlled by the syringe pumps. In the wet chemistry procedure, the titration of the asphaltenes is well beyond 1:40. In fact, the ratio is not exactly known. The Soxhlet washing step continues until the solvent is clear, which is an unquantifiable step. Therefore, the separated asphaltenes in the two techniques are not identical. Furthermore, the Soxhlet washing step takes place with hot, nearly boiling, heptane. In the microfluidic technique, the precipitation occurs at 30°C, which results in precipitation of a slightly different class of molecules.

Here, we correlate the spectrophotometry measurements to asphaltene contents generated using a particular wet chemistry method. Other wet chemistry procedures can also be used as the calibration standard. One may envisage using other titrants such as pentane or hexane, different temperatures, and/or other procedures, which could potentially lead to a different and perhaps a better correlation. Further investigation is required to improve the correlation quality. The basic microfluidic device is clearly capable of solving a variety of customized asphaltene separation and quantitation measurements.

The data reported by Sieben et al. [100] suggest that the variation in molar absorptivity is reasonably small for asphaltene distributions that span a wide range of crude oils. Considering the diversity of the crude samples and the sources of errors that discussed, the plot shows a good global correlation for characterization and quantification of asphaltenes.

The limit of detection and the limit of quantification of the technique have been briefly studied and are largely limited by (i) the optical path length selected, (ii) the detector used to measure absorbance, and (iii) fluid pulsations and mixing. In the described system, the combined effects led to a standard deviation on a blank measurement of typically ±0.005 au. Taking the somewhat conventional definition for the limit of detection (LOD) as 3 times the standard deviation and for limit of quantitation (LOQ) as 10 times the standard deviation yields an LOD of ±0.1 wt.% (0.015 au) and an LOQ of 0.3 wt.% (0.05 au). To measure lower asphaltene concentrations, the optical path length can be increased from millimeters to centimeters in scale, as per the Beer–Lambert law.

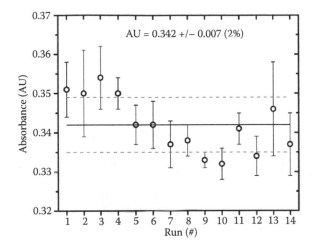

FIGURE 11.18 Optical absorption results for 14 measurements of the same crude oil sample highlighting the reproducibility of the system. (Reprinted with permission from Schneider, M. H. et al., Measurement of asphaltenes using optical spectroscopy on a microfluidic platform, *Anal. Chem., 85* (10), 5153–5160. Copyright 2013 American Chemical Society.)

11.6.2 REPEATABILITY

To investigate the repeatability of the microfluidic method, the apparatus was evaluated using a stable crude sample for several months. The sample had an asphaltene content of 1.1 wt.% with an average absorbance of 0.34. Figure 11.18 depicts the absorbance of asphaltenes for the sample measured 14 times during a 3-month period. The measurements showed that the variations were rather small. The relative standard deviation was ±2%, and all measurements fell within ±3% of the average. Considering the large errors inherent in wet chemical methods for characterizing low-asphaltene-content fluids (50% for <1 wt.% asphaltene content and 20% for fluids with an asphaltene content between 1 and 3 wt.%), the microfluidic technique shows a remarkable repeatability. It seems likely that such methods may prove to be a more reliable approach to producing high-quality accurate and precise data than the traditional approaches to asphaltene separation and assessment.

11.6.3 REDISSOLUTION

The previous data indirectly measured the asphaltene spectrum by subtracting the maltene spectrum from the oil spectrum. Alternatively, the spectrum of the asphaltenes can be directly measured by redissolving the deposited material on the filter membrane using toluene or dichloromethane; this is shown in Figure 11.19. This process starts similarly to the spectral difference approach in that the initial absorbance of diluted oil is measured. Asphaltenes are then precipitated, and the absorbance of the filtered maltenes is measured. Finally, the deposited asphaltenes are liberated and the optical absorbance versus time is recorded as the asphaltenes

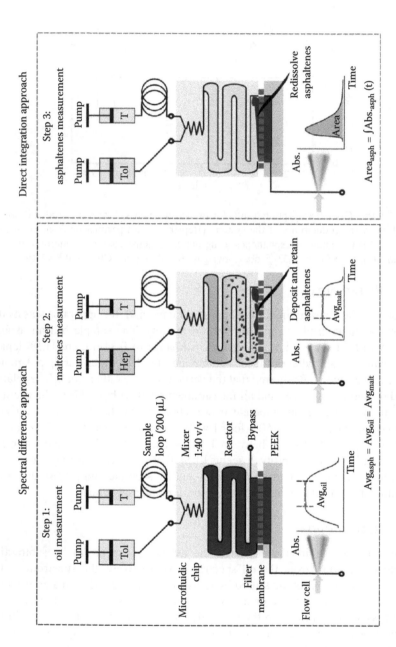

FIGURE 11.19 Two microfluidic approaches for measuring asphaltene optical absorbance and for determining asphaltene content in crude oil.

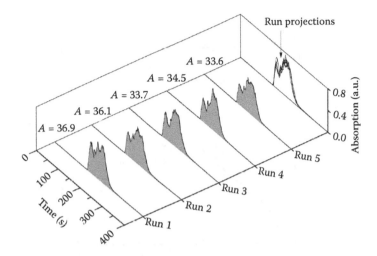

FIGURE 11.20 Consecutive redissolutions of collected asphaltenes. In total, five experiments show the repeatable nature of dissolution by absorption area and peak shape. (Copyright 2013, Society of Petroleum Engineers Inc. Reproduced with permission of SPE. Further reproduction prohibited without permission.)

are redissolved. The signal is then integrated to determine the total absorbance area, which should correlate to the total asphaltene mass content.

When the system was used in this mode, higher-asphaltene-content oils dissolved too rapidly and produced absorbance values >3 (0.1% light transmission), which is beyond the dynamic range of the spectrometer. Therefore, the rapid release of asphaltenes led to the signal exceeding the dynamic range of the optical system. Creating a controlled release of deposited asphaltenes would require redesign of the microfluidic chip or reconfiguration of the optical system. Reducing the amount of deposited asphaltenes would also prevent saturation of the optical system. To this end, the sample loop volume was reduced from 200 to 5 µL. The experiments were repeated and the signal was indeed within tolerable ranges of <2 absorption units.

Figure 11.20 shows the redissolved asphaltene optical absorbance signals for a single oil sample—data shown are from five experiments. Both the plot profiles and the integrated absorption areas, shown above each plot, demonstrate the repeatability of the measurement. The signature optical profile was unique for each oil sample and is likely the product of system fluid dynamics and varied solubility of asphaltene molecules comprising the sample.

With the ability to directly measure the asphaltene absorbance area, 16 different oil samples of varying asphaltene content were evaluated (plotted in Figure 11.21). The data show the optical absorbance area versus the weight percentage of the asphaltenes. The relationship is linear with a similar degree of scatter, as was present in the spectral difference correlation. The absorbance area of asphaltenes was then plotted versus the average absorbance difference between the oil and maltenes, as per the spectral difference method. The strong correlation with minimal scatter suggests that there is optical conservation. This, in turn, validates the difference measurement as an indirect method of determining asphaltene content. It should be noted that a

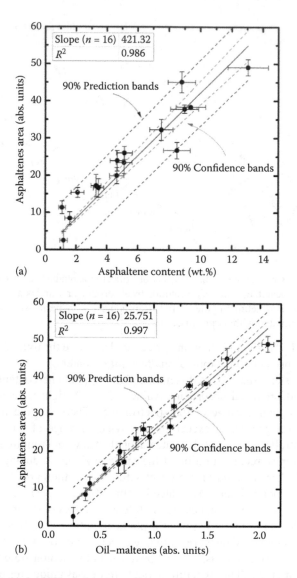

FIGURE 11.21 (a) Asphaltene absorbance areas versus mass measurements. (b) Asphaltene absorbance areas versus absorbance differences of oils and maltenes. Optical conservation between the oil, maltene, and asphaltene spectra is demonstrated by the linear correlation. (Copyright 2013, Society of Petroleum Engineers Inc. Reproduced with permission of SPE. Further reproduction prohibited without permission.)

few samples would leave residual asphaltenes on the filter membrane. The literature indicates that the forward and reverse reaction rates for precipitation and dissolution may occur on different time scales [101–103]. In this approach, the small portions of asphaltenes that do not redissolve are not accounted for in the absorbance integration measurement and could introduce minor errors in the measurements.

11.7 CONCLUSIONS

In this chapter, we discussed the emergence of a new and exciting analytical technology in the oil field industry. We presented a microfluidic approach to asphaltene content measurement. In that technique, the conventional precipitation procedure using a titrant was miniaturized, while a UV–Vis spectrometry method was incorporated for detection and quantitation. The fundamentals of the microfluidic technique including fluid mechanics, optical absorption spectroscopy, and dispersion were discussed in detail.

The method required only 200 µL of sample with a few milliliters of solvents. A chaotic mixer mixed the crude sample with n-heptane at a 1:40 v/v ratio. The precipitated asphaltenes were passed through a 200-nm-pore membrane. The optical spectrum of the maltenes was collected downstream of the membrane. The optical spectrum of the asphaltenes was indirectly measured by subtracting the optical spectrum of the maltenes from that of the crude oil. We showed a strong correlation between the optical spectrum acquired by this technique and the asphaltene content by using the conventional wet chemistry method. The method showed an excellent repeatability during a 3-month period.

Furthermore, we showed a similar correlation between asphaltene content and the optical spectrum when the spectrum of asphaltenes is measured directly by redissolving the asphaltenes. In that technique, the asphaltenes were precipitated on the membrane similar to the previous protocol. However, instead of calculating the difference in the optical spectra of oil and maltenes, the collected asphaltenes on the membrane were redissolved using toluene. As the asphaltenes were redissolved, the temporal response of the spectrometer was collected. The total integrated area under the curve correlated well with the quantities obtained using the subtraction technique.

The potentials of microfluidics and the lab-on-a-chip concept in the oil and gas industry are enormous. We believe that by incorporating this technology into the industry, valuable information and data can be generated at a much faster pace, and it can be of a higher quality. This work is an example of such potential, which paves the road for future applications.

ACKNOWLEDGMENTS

We acknowledge Drs. John Ratulowski, Oliver Mullins, Artur Stankiewicz, Shahnawaz Molla, and Steve Larter for fruitful discussions.

REFERENCES

1. Mullins, O. C.; Sheu, E. Y.; Hammami, A.; Marshall, A. G., *Asphaltenes, Heavy Oils and Petroleomics*. Springer: New York, 2007.
2. Mullins, O. C.; Sabbah, H.; Eyssautier, J.; Pomerantz, A. E.; Barre, L.; Andrews, A. B.; Ruiz-Morales, Y.; Mostowfi, F.; McFarlane, R.; Goual, L.; Lepkowicz, R.; Cooper, T.; Orbulescu, J.; Leblanc, R. M.; Edwards, J.; Zare, R. N., Advances in asphaltene science and the Yen–Mullins model. *Energy Fuels* **2012**, *26* (7), 3986–4003.
3. Wilhelms, A.; Larter, S. R., Origin of tar mats in petroleum reservoirs. 1. Introduction and case-studies. *Mar. Pet. Geol.* **1994**, *11* (4), 418–441.

4. Wilhelms, A.; Larter, S. R., Origin of tar mats in petroleum reservoirs. 2. Formation mechanisms for tar mats. *Mar. Pet. Geol.* **1994,** *11* (4), 442–456.
5. Alshareef, A. H.; Scherer, A.; Tan, X.; Azyat, K.; Stryker, J. M.; Tykwinski, R. R.; Gray, M. R., Formation of archipelago structures during thermal cracking implicates a chemical mechanism for the formation of petroleum asphaltenes. *Energy Fuels* **2011,** *25* (5), 2130–2136.
6. Borton, D., II; Pinkston, D. S.; Hurt, M. R.; Tan, X.; Azyat, K.; Scherer, A.; Tykwinski, R.; Gray, M.; Qian, K.; Kenttaemaa, H. I., Molecular structures of asphaltenes based on the dissociation reactions of their ions in mass spectrometry. *Energy Fuels* **2010,** *24,* 5548–5559.
7. Mostowfi, F.; Indo, K.; Mullins, O. C.; McFarlane, R., Asphaltene nanoaggregates studied by centrifugation. *Energy Fuels* **2009,** *23,* 1194–1200.
8. Gray, M. R.; Tykwinski, R. R.; Stryker, J. M.; Tan, X., Supramolecular assembly model for aggregation of petroleum asphaltenes. *Energy Fuels* **2011,** *25* (7), 3125–3134.
9. Mostowfi, F.; Czarnecki, J.; Masliyah, J.; Bhattacharjee, S., A microfluidic electrochemical detection technique for assessing stability of thin films and microemulsions. *J. Colloid Interface Sci.* **2008,** *317* (2), 593.
10. Di Primio, R.; Horsfield, B.; Guzman-Vega, M. A., Determining the temperature of petroleum formation from the kinetic properties of petroleum asphaltenes. *Nature* **2000,** *406* (6792), 173–176.
11. Akbarzadeh, K.; Dhillon, A.; Svrcek, W. Y.; Yarranton, H. W., Methodology for the characterization and modeling of asphaltene precipitation from heavy oils diluted with *n*-alkanes. *Energy Fuels* **2004,** *18* (5), 1434–1441.
12. Alboudwarej, H.; Beck, J.; Svrcek, W. Y.; Yarranton, H. W.; Akbarzadeh, K., Sensitivity of asphaltene properties to separation techniques. *Energy Fuels* **2002,** *16* (2), 462–469.
13. Andersen, S. I., Effect of precipitation temperature on the composition of n-heptane asphaltenes. *Fuel Sci. Technol. Int.* **1994,** *12* (1), 51–74.
14. Andersen, S. I.; Keul, A.; Stenby, E., Variation in composition of subfractions of petroleum asphaltenes. *Pet. Sci. Technol.* **1997,** *15* (7–8), 611–645.
15. Buckley, J. S.; Hirasaki, G. J.; Liu, Y.; Von Drasek, S.; Wang, J. X.; Gill, B. S., Asphaltene precipitation and solvent properties of crude oils. *Pet. Sci. Technol.* **1998,** *16* (3–4), 251–285.
16. Hirschberg, A.; deJong, L. N. J.; Schipper, B. A.; Meijer, J. G., Influence of temperature and pressure on asphaltene flocculation. *Soc. Pet. Eng. J.* **1984,** *24* (3), 283–293.
17. Katz, D. L.; Beu, K. E., Nature of asphaltic substances. *Ind. Eng. Chem.* **1945,** *37* (2), 195–200.
18. Mansoori, G. A.; Vazquez, D.; Shariaty-Niassar, M., Polydispersity of heavy organics in crude oils and their role in oil well fouling. *J. Pet. Sci. Eng.* **2007,** *58* (3–4), 375–390.
19. Rogel, E.; Ovalles, C.; Moir, M., Asphaltene chemical characterization as a function of solubility: Effects on stability and aggregation. *Energy Fuels* **2012,** *26* (5), 2655–2662.
20. Sabbagh, O.; Akbarzadeh, K.; Badamchi-Zadeh, A.; Svrcek, W. Y.; Yarranton, H. W., Applying the PR-EoS to asphaltene precipitation from *n*-alkane diluted heavy oils and bitumens. *Energy Fuels* **2006,** *20* (2), 625–634.
21. Zhao, B.; Shaw, J. M., Composition and size distribution of coherent nanostructures in Athabasca bitumen and Maya crude oil. *Energy Fuels* **2007,** *21* (5), 2795–2804.
22. Sheu, E. Y., Petroleum asphaltene properties, characterization, and issues. *Energy Fuels* **2002,** *16* (1), 74–82.
23. Wattana, P.; Fogler, H. S.; Yen, A.; Carmen Garcìa, M. D.; Carbognani, L., Characterization of polarity-based asphaltene subfractions. *Energy Fuels* **2005,** *19* (1), 101–110.
24. Yarranton, H. W.; Masliyah, J. H., Molar mass distribution and solubility modeling of asphaltenes. *AIChE J.* **1996,** *42* (12), 3533–3543.

25. Maqbool, T.; Balgoa, A. T.; Fogler, H. S., Revisiting asphaltene precipitation from crude oils: A case of neglected kinetic effects. *Energy Fuels* **2009,** *23* (7), 3681–3686.
26. American Society for Testing and Materials (ASTM), *ASTM D6560, Standard Test Method for Determination of Asphaltenes (Heptane Insolubles) in Crude Petroleum and Petroleum Products.* ASTM: West Conshohocken, PA, 2000.
27. Ríos, Á.; Zougagh, M.; Avila, M., Miniaturization through lab-on-a-chip: Utopia or reality for routine laboratories? A review. *Anal. Chim. Acta* **2012,** *740,* 1–11.
28. Whitesides, G. M., The origins and the future of microfluidics. *Nature* **2006,** *442* (7101), 368–373.
29. Manz, A.; Graber, N.; Widmer, H. M., Miniaturized total chemical analysis systems: A novel concept for chemical sensing. *Sens. Actuators B Chem.* **1990,** *1* (1–6), 244–248.
30. Harrison, D. J.; Fluri, K.; Seiler, K.; Fan, Z. H.; Effenhauser, C. S.; Manz, A., Micromachining a miniaturized capillary electrophoresis-based chemical-analysis system on a chip. *Science* **1993,** *261* (5123), 895–897.
31. Zhang, C. S.; Xu, J. L.; Ma, W. L.; Zheng, W. L., PCR microfluidic devices for DNA amplification. *Biotechnol. Adv.* **2006,** *24* (3), 243–284.
32. Beebe, D. J.; Mensing, G. A.; Walker, G. M., Physics and applications of microfluidics in biology. *Annu. Rev. Biomed. Eng.* **2002,** *4,* 261–286.
33. Sanders, G. H. W.; Manz, A., Chip-based microsystems for genomic and proteomic analysis. *TrAC, Trends Anal. Chem.* **2000,** *19* (6), 364–378.
34. Situma, C.; Hashimoto, M.; Soper, S. A., Merging microfluidics with microarray-based bioassays. *Biomol. Eng.* **2006,** *23* (5), 213–231.
35. Lynch, M.; Mosher, C.; Huff, J.; Nettikadan, S.; Johnson, J.; Henderson, E., Functional protein nanoarrays for biomarker profiling. *Proteomics* **2004,** *4* (6), 1695–1702.
36. Zare, R. N.; Kim, S., Microfluidic platform for single-cell analysis. *Ann. Rev. Biomed. Eng.* **2010,** *12,* 187–201.
37. Paguirigan, A. L.; Beebe, D. J., Microfluidics meet cell biology: Bridging the gap by validation and application of microscale techniques for cell biological assays. *Bioessays* **2008,** *30* (9), 811–821.
38. Andersson, H.; van den Berg, A., Microfabrication and microfluidics for tissue engineering: State of the art and future opportunities. *Lab Chip* **2004,** *4* (2), 98–103.
39. Khademhosseini, A.; Langer, R.; Borenstein, J.; Vacanti, J. P., Microscale technologies for tissue engineering and biology. *Proc. Natl. Acad. Sci. U. S. A.* **2006,** *103* (8), 2480–2487.
40. Weigl, B. H.; Bardell, R. L.; Cabrera, C. R., Lab-on-a-chip for drug development. *Adv. Drug Deliv. Rev.* **2003,** *55* (3), 349–377.
41. Neuzil, P.; Giselbrecht, S.; Lange, K.; Huang, T. J.; Manz, A., Revisiting lab-on-a-chip technology for drug discovery. *Nat. Rev. Drug Discov.* **2012,** *11* (8), 620–632.
42. Choban, E. R.; Markoski, L. J.; Wieckowski, A.; Kenis, P. J. A., Microfluidic fuel cell based on laminar flow. *J. Power Sources* **2004,** *128* (1), 54–60.
43. Kjeang, E.; Djilali, N.; Sinton, D., Microfluidic fuel cells: A review. *J. Power Sources* **2009,** *186* (2), 353–369.
44. Togo, M.; Takamura, A.; Asai, T.; Kaji, H.; Nishizawa, M., Structural studies of enzyme-based microfluidic biofuel cells. *J. Power Sources* **2008,** *178* (1), 53–58.
45. Mauriz, E.; Calle, A.; Manclus, J. J.; Montoya, A.; Lechuga, L. M., Multi-analyte SPR immunoassays for environmental biosensing of pesticides. *Anal. Bioanal. Chem.* **2007,** *387* (4), 1449–1458.
46. Marle, L.; Greenway, G. M., Microfluidic devices for environmental monitoring. *TrAC, Trends Anal. Chem.* **2005,** *24* (9), 795–802.
47. Herold, K. E.; Rasooly, A., *Lab on a Chip Technology Volume 1: Fabrication and Microfluidics,* Vol. 1. Caister Academic Press: Norfolk, UK, 2009.

48. Jackson, M. J., *Microfabrication and Nanomanufacturing*. Taylor & Francis Group: Boca Raton, FL, 2006.
49. Madou, M. J., *Fundamentals of Microfabrication: The Science of Miniaturization, Second Edition*. Taylor & Francis: Boca Raton, FL, 2002.
50. Li, D. (Ed.), *Encyclopedia of Microfluidics and Nanofluidics*. Springer: New York, 2008.
51. Tantra, R.; van Heeren, H., Product qualification: A barrier to point-of-care microfluidic-based diagnostics? *Lab Chip* **2013**, *13* (12), 2199–2201.
52. Mattax, C. C.; Kyte, J. R., Ever see a water flood. *Oil Gas J.* **1961**, *59* (42), 115–128.
53. Bonnet, J.; Lenormand, R., Constructing micromodels for study of multiphase flow in porous-media. *Rev. Inst. Francais Pet.* **1977**, *32* (3), 477–480.
54. Chang, L. C.; Tsai, J. P.; Shan, H. Y.; Chen, H. H., Experimental study on imbibition displacement mechanisms of two-phase fluid using micro model. *Environ. Earth Sci.* **2009**, *59* (4), 901–911.
55. van Dijke, M. I. J.; Sorbie, K. S.; Sohrabi, A.; Danesh, A., Simulation of WAG floods in an oil-wet micromodel using a 2-D pore-scale network model. *J. Pet. Sci. Eng.* **2006**, *52* (1–4), 71–86.
56. Mostowfi, F.; Khristov, K.; Czarnecki, J.; Masliyah, J.; Bhattacharjee, S., Electric field mediated breakdown of thin liquid films separating microscopic emulsion droplets. *Appl. Phys. Lett.* **2007**, *90* (18), 184102.
57. Kharrat, A. M.; Indo, K.; Mostowfi, F., Asphaltene content measurement using an optical spectroscopy technique. *Energy Fuels* **2013**, *27* (5), 2452–2457.
58. Schneider, M. H.; Sieben, V. J.; Kharrat, A. M.; Mostowfi, F., Measurement of asphaltenes using optical spectroscopy on a microfluidic platform. *Anal. Chem.* **2013**, *85* (10), 5153–5160.
59. Bowden, S. A.; Monaghan, P. B.; Wilson, R.; Parnell, J.; Cooper, J. M., The liquid–liquid diffusive extraction of hydrocarbons from a North Sea oil using a microfluidic format. *Lab Chip* **2006**, *6* (6), 740–743.
60. Bowden, S. A.; Wilson, R.; Parnell, J.; Cooper, J. M., Determination of the asphaltene and carboxylic acid content of a heavy oil using a microfluidic device. *Lab Chip* **2009**, *9* (6), 828–832.
61. Abolhasani, M.; Singh, M.; Kumacheva, E.; Günther, A., Automated microfluidic platform for studies of carbon dioxide dissolution and solubility in physical solvents. *Lab Chip* **2012**, *12* (9), 1611–1618.
62. Lefortier, S. G. R.; Hamersma, P. J.; Bardow, A.; Kreutzer, M. T., Rapid microfluidic screening of CO_2 solubility and diffusion in pure and mixed solvents. *Lab Chip* **2012**, *12* (18), 3387–3391.
63. Fadaei, H.; Scarff, B.; Sinton, D., Rapid microfluidics-based measurement of CO_2 diffusivity in bitumen. *Energy Fuels* **2011**, *25* (10), 4829–4835.
64. Fadaei, H.; Shaw, J. M.; Sinton, D., Bitumen–toluene mutual diffusion coefficients using microfluidics. *Energy Fuels* **2013**, *27* (4), 2042–2048.
65. Mostowfi, F.; Molla, S.; Tabeling, P., Determining phase diagrams of gas–liquid systems using a microfluidic PVT. *Lab Chip* **2012**, *12* (21), 4381–4387.
66. Fisher, R.; Shah, M. K.; Eskin, D.; Schmidt, K.; Singh, A.; Molla, S.; Mostowfi, F., Equilibrium gas–oil ratio measurements using microfluidic technique. *Lab Chip* **2013**, *13* (13), 2623–2633.
67. Nguyen, N. T.; Wu, Z., Micromixers—A review. *J. Micromech. Microeng.* **2005**, *15* (2), R1–R16.
68. Unger, M. A.; Chou, H. P.; Thorsen, T.; Scherer, A.; Quake, S. R., Monolithic microfabricated valves and pumps by multilayer soft lithography. *Science* **2000**, *288* (5463), 113–116.
69. Floquet, C. F. A.; Sieben, V. J.; Milani, A.; Joly, E. P.; Ogilvie, I. R. G.; Morgan, H.; Mowlem, M. C., Nanomolar detection with high sensitivity microfluidic absorption cells manufactured in tinted PMMA for chemical analysis. *Talanta* **2011**, *84* (1), 235–239.

70. Sieben, V. J.; Floquet, C. F. A.; Ogilvie, I. R. G.; Mowlem, M. C.; Morgan, H., Microfluidic colourimetric chemical analysis system: Application to nitrite detection. *Anal. Methods* **2010**, *2* (5), 484–491.

71. Behnam, M.; Kaigala, G. V.; Khorasani, M.; Martel, S.; Elliott, D. G.; Backhouse, C. J., Integrated circuit-based instrumentation for microchip capillary electrophoresis. *IET Nanobiotechnol.* **2010**, *4* (3), 91–101.

72. Ruiz-Morales, Y.; Wu, X.; Mullins, O. C., Electronic absorption edge of crude oils and asphaltenes analyzed by molecular orbital calculations with optical spectroscopy. *Energy Fuels* **2007**, *21* (2), 944–952.

73. Hecht, E., *Optics*, 4th ed. Addison Wesley: San Francisco, 2002.

74. Sciau, P., Nanoparticles in ancient materials: The metallic lustre decorations of medieval ceramics. In *The Delivery of Nanoparticles*, Hashim, A. A., (Ed.), InTech: Rijeka, Croatia, 2012.

75. Atkins, P. W.; Friedman, R. S., *Molecular Quantum Mechanics*. OUP: Oxford, UK, 2011.

76. Costner, E. A.; Long, B. K.; Navar, C.; Jockusch, S.; Lei, X.; Zimmerman, P.; Campion, A.; Turro, N. J.; Willson, C. G., Fundamental optical properties of linear and cyclic alkanes: VUV absorbance and index of refraction. *J. Phys. Chem. A* **2009**, *113* (33), 9337–9347.

77. Shimadzu, The relationship between UV–VIS absorption and structure of organic compounds. *UV Talk Lett.* **2009**, *2*, 1–8.

78. Shevell, S. K. (ed.), *The Science of Color*, 2nd ed. Elsevier: Amsterdam, Boston, Optical Society of America, 2003.

79. Friedel, R. A.; Orchin, M., Ultraviolet spectra of aromatic compounds, Wiley: New York, 1951.

80. Speight, J. G., *The Chemistry and Technology of Petroleum*. Marcel Dekker: New York, 1999.

81. Zhao, B.; Becerra, M.; Shaw, J. M., On asphaltene and resin association in Athabasca bitumen and Maya crude oil. *Energy Fuels* **2009**, *23* (9), 4431–4437.

82. Derakhshesh, M.; Gray, M. R.; Dechaine, G. P., Dispersion of asphaltene nanoaggregates and the role of Rayleigh scattering in the absorption of visible electromagnetic radiation by these nanoaggregates. *Energy Fuels* **2013**, *27* (2), 680–693.

83. Mansoori, G. A., A unified perspective on the phase behaviour of petroleum fluids. *Int. J. Oil Gas Coal Technol.* **2009**, *2* (2), 141–167.

84. Ruiz-Morales, Y.; Mullins, O. C., Polycyclic aromatic hydrocarbons of asphaltenes analyzed by molecular orbital calculations with optical spectroscopy. *Energy Fuels* **2007**, *21* (1), 256–265.

85. Bouquet, M.; Hamon, J. Y., Determination of asphaltene content in petroleum products for concentrations below 20000 ppm down to 150 ppm. *Fuel* **1985**, *64* (11), 1625–1627.

86. Kirby, B., J., *Micro- and Nanoscale Fluid Mechanics—Transport in Microfluidic Devices*. Cambridge University Press: New York, 2010.

87. Steinke, M. E.; Kandlikar, S. G., Single-phase liquid friction factors in microchannels. *Int. J. Therm. Sci.* **2006**, *45* (11), 1073–1083.

88. Kandlikar, S. G.; Garimella, S.; Li, D.; Colin, S.; King, M. R., *Heat Transfer and Fluid Flow in Minichannels and Microchannels*, 1st ed. Elsevier: Amsterdam, 2006.

89. Sharp, K. V.; Adrian, R. J.; Santiago, J. G.; Molho, J. I., Liquid flows in microchannels. In *The MEMS Handbook*, Gad-el-Hak, M., Ed. CRC Press: Boca Raton, FL, 2002.

90. Molla, S.; Eskin, D.; Mostowfi, F., Pressure drop of slug flow in microchannels with increasing void fraction: Experiment and modeling. *Lab Chip* **2011**, *11* (11), 1968–1978.

91. Molla, S.; Eskin, D.; Mostowfi, F., Two-phase flow in microchannels: The case of binary mixtures. *Ind. Eng. Chem. Res.* **2013**, *52*, 941–953.

92. White, F., M., *Viscous Fluid Flow*, 3rd ed. McGraw Hill: New York, 2006.

93. Stone, H. A.; Stroock, A. D.; Ajdari, A., Engineering flows in small devices: Microfluidics toward a lab-on-a-chip. *Annu. Rev. Fluid Mech.* **2004**, *36*, 381–411.

94. Ottino, J. M.; Wiggins, S., Introduction: Mixing in microfluidics. *Philos. Trans. R. Soc. A Math. Phys. Eng. Sci.* **2004,** *362* (1818), 923–935.

95. Stroock, A. D.; Dertinger, S. K. W.; Ajdari, A.; Mezić, I.; Stone, H. A.; Whitesides, G. M., Chaotic mixer for microchannels. *Science* **2002,** *295* (5555), 647–651.

96. Taylor, G., Dispersion of soluble matter in solvent flowing slowly through a tube. *Proc. R. Soc. Lond. Ser. A Math. Phys. Sci.* **1953,** *219* (1137), 186–203.

97. Aris, R., On the dispersion of a solute in a fluid flowing through a tube. *Proc. R. Soc. Lond. Ser. A Math. Phys. Sci.* **1956,** *235* (1200), 67–77.

98. Bahrami, M.; Yovanovich, M. M.; Culham, J. R., Pressure drop of fully-developed, laminar flow in microchannel of arbitrary cross-section. *J. Fluids Eng. Trans. ASME* **2006,** *128* (5), 1036–1044.

99. Schneider, M. H.; Andrews, A. B.; Mitra-Kirtley, S.; Mullins, O. C., Asphaltene molecular size by fluorescence correlation spectroscopy. *Energy Fuels* **2007,** *21*, 2875–2882.

100. Sieben, V. J.; Kharrat, A. M.; Mostowfi, F., Novel measurement of asphaltene content in oil using microfluidic technology. In *SPE Annual Technical Conference and Exhibition*, Vol. SPE-166394-MS. Society of Petroleum Engineers, New Orleans, LA, 2013.

101. Abedini, A.; Ashoori, S.; Torabi, F.; Saki, Y.; Dinarvand, N., Mechanism of the reversibility of asphaltene precipitation in crude oil. *J. Pet. Sci. Eng.* **2011,** *78* (2), 316–320.

102. Beck, J.; Svrcek, W. Y.; Yarranton, H. W., Hysteresis in asphaltene precipitation and redissolution. *Energy Fuels* **2005,** *19* (3), 944–947.

103. Andersen, S. I.; Stenby, E., II, Thermodynamics of asphaltene precipitation and dissolution investigation of temperature and solvent effects. *Fuel Sci. Technol. Int.* **1996,** *14* (1–2), 261–287.

Section VI

Modeling and Chemometrics

12 Application of Data Fusion for Enhanced Understanding and Control

Thomas I. Dearing, Rachel Mohler,
Carl E. Rechsteiner Jr., and Brian J. Marquardt

CONTENTS

CONTEXT

Chemometrics is widely used to infer properties on the basis of spectral and physical measurements. The challenge for accurate property prediction is aided by data fusion, which

- Merges multiple measurements to reduce the error of prediction
- Minimizes bias due to a single measurement through proper data scaling
- Has multiple levels to minimize the computation time and related efforts for a good prediction

ABSTRACT

During the past couple of decades, there has been a series of paradigm shifts moving laboratory-based instrumentation out onto process streams for process control.

This laboratory instrumentation underwent a number of evolutions that resulted in the generation of process analyzers. Spectroscopic methods such as Raman, near infrared, infrared, and nuclear magnetic resonance have become increasingly popular within the oil and gas industries, owing to their rapid analysis cycles and the fidelity of data they provide. Gas chromatography (GC) also underwent stages of process readiness and miniaturization. GC is particularly suited to petrochemical analysis, owing to its ability to split complex mixtures based on the boiling points of the individual constituents. Process-ready Micro-Fast GCs are able to process fuel samples in a fraction of the time when compared with laboratory instruments. The ability to collect data in real time is only half of the solution to the process control challenge. The data generated by process analyzers must be examined in real time to generate information that can be used to control a process. The field of chemometrics has long been established as the main set of techniques for the analysis of process data. The coupling of chemometrics with process analyzers unlocked a large volume of information regarding processes. The unlocked information and the potential to unlock further information about processes drove a desire to get more analyzers to the process streams.

The proliferation of analytical measurements at specific points in a process creates a unique opportunity to combine the measurements made by all of the analyzers to more effectively map the process data space and improve the control and understanding of the process. Data fusion is a subclass of chemometrics procedures that combine data or models into one contiguous fused entity. The value of data fusion can be attributed to the ability to fuse together different sources of information that may be measuring different parts of a process. This chapter will outline the main methods of performing data fusion and the interesting results from two specific petrochemical case studies.

12.1 INTRODUCTION

During the past couple of decades, there has been a series of paradigm shifts moving laboratory-based instrumentation out onto process streams for process control. This laboratory instrumentation underwent a series of evolutions that resulted in the generation of process analyzers. Spectroscopic methods such as Raman, near-infrared, infrared (IR), and nuclear magnetic resonance (NMR) have become increasingly popular owing to their rapid analysis cycles and the fidelity of the data they provide. Measurements using IR-based systems have increased dramatically, in particular in the areas of fuel and oil refining [1,2]. This increase can be attributed to IR offering fast and nondestructive vibrational spectroscopy measurements, which, when coupled with immersion probes, can perform online real-time analysis that are rich in data for modeling and well suited to petrochemical high-value products. The use of Raman spectroscopy within fuel and oil refining has also seen a steady increase in recent years. Raman spectroscopy offers complementary information to IR-based systems, and by using advanced probe technology can provide an even greater amount of chemical information when examining certain oil and fuel distillation fractions. Until recent years, NMR had been solely reserved for laboratory analysis; however, the developments in rare earth magnets and miniaturized RF coils had enabled NMR to move to the process line. NMR spectral measurements provide information regarding the electronic structure and conformational arrangement of a molecule. Within a petrochemical context, this allows the elucidation of a fuel's composition in terms of aliphatic and aromatic molecules. Process NMR instruments typically employ magnets that operate in the range of 60 MHz; the free induction decay (FID) curves are captured and transformed using a Fourier transform (FT) to yield NMR spectra that can be used for process control and optimization [3,4]. Gas chromatography (GC) also underwent stages of process readiness and miniaturization. GC is particularly suited to petrochemical analysis, because of its ability to split complex mixtures based on the boiling points of the individual constituents. Laboratory-based GCs are typically able to characterize a fuel sample in approximately 30–120 min. The process-ready Micro-Fast GCs are able to process the same fuel sample in only 3–10 min. The massive reduction of cycle time facilitates the movement of GC to a process stream for real-time analysis. The ability to collect data in real time is only half of the process control challenge.

The data generated by process analyzers must be examined in real time to generate information that can be used to control a process. The field of chemometrics has long been established as the main set of techniques for the analysis of process data. Championed by Bruce Kowalski at the Center for Process Analytical Chemistry, chemometrics was exposed to a wide array of industrial applications. Techniques such as principal component analysis (PCA) and partial least squares (PLS) were used to solve a number of process problems. Typically, chemometrics is applied in real time through the use of calibration and prediction models. Calibration models are built offline using spectra collected from the process and laboratory data related to the chosen prediction property. The performance of the calibration model is determined through the root mean square errors in calibration and cross validation. The calibration model is vetted through the prediction of known test data to ensure model accuracy and reproducibility. The performance of the prediction model is measured by the root mean square error of prediction (RMSEP). The prediction model is then deployed online to predict the chosen property in real time using the newly measured spectra. At regular intervals, model maintenance will be performed to guarantee continued model performance.

The adoption of chemometrics grew within the petrochemical industry, using the data generated by process analyzers to build models that were able to determine the number of high-value process descriptors [5–7]. The coupling of chemometrics with process analyzers unlocked a large volume of information regarding processes. The unlocked information and the potential to unlock further information about processes drove a desire to bring more analyzers to the process streams. This required the development of a reproducible and flexible sampling interface. NeSSI (New Sampling/Sensor Initiative) sampling systems are combinations of modular components for physical fluid handling. An example of a NeSSI sampling system is shown in Figure 12.1 (see Chapter 2 for a discussion on the NeSSI sampling system). Use of NeSSI components that were specifically designed to accept probes and analytical sensors provided the means to bring a large array of analytics to a specific location of a process stream.

The proliferation of analytical measurements at specific points in a process creates a unique opportunity to combine the measurements made by all of the analyzers to more effectively map the process data space and improve the control and understanding of the process. Data fusion is a subclass of chemometrics procedures that combine

FIGURE 12.1 NeSSI sampling system.

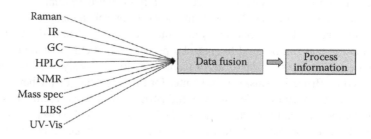

FIGURE 12.2 Data fusion concept flow chart.

data or models into one contiguous fused entity. This is illustrated in Figure 12.2. The value of data fusion can be attributed to the ability to fuse together different sources of information that may be measuring different parts of a process. The principle of adding different, orthogonal information to a data fused model is outlined in Figures 12.3 through 12.5. Figure 12.3 shows images of an object collected along a complimentary axis. The images have minor differences that when fused together would yield additional information regarding the object. At this stage, the object could probably be identified; however, little more would be deduced. A third image can be collected along an orthogonal axis (Figure 12.4). By fusing together the two complimentary images with the orthogonal image to yield a fused image (Figure 12.5), significantly more information can be derived that would not have been present if the images were observed individually. There are numerous forms of data fusion, all of which can be subdivided into three levels: low, medium, and high.

FIGURE 12.3 Complimentary measurements made of an object.

FIGURE 12.4   Orthogonal measurement of an object.

FIGURE 12.5   Fused measurements of the two complimentary and one orthogonal images.

12.1.1   Low-Level

Low-level data fusion methods begin by processing the raw data to remove any variation not associated with an analytical response. Next, the processed data are normalized, either to unit area or until height and scaled to unit length. The normalized data are then concatenated to form a continuous signal. The fused data is then processed and analyzed using standard chemometric methods such as PCA and PLS. Low-level data fusion is simple to apply, owing to the need for one calibration model built from the fused data. The key advantage for using low-level data fusion is the necessity to build only one calibration model, making this approach the simplest of the three levels of data fusion. The main disadvantage of using low-level data fusion is the use of concatenation. By using concatenation for fusion, careful normalization and scaling must be performed to prevent the data from one analyzer from overwhelming the other data sets. In addition, by concatenating the data from each of the process analyzers, the fused data matrix has a length that is equal to the sum of the lengths of the data from each of the individual analyzers. The resulting large data matrix can cause significant increases in the time needed to produce the calibration and prediction models.

12.1.2   Medium-Level

Medium-level data fusion (see Figure 12.6) begins by processing the individual process analyzer data to remove unwanted variation. The processed data is then reduced (using PCA or PLS) to its scores, generating a set of scores for each data set. The scores

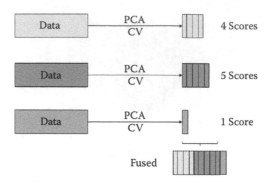

FIGURE 12.6   Medium-level data fusion.

of a matrix relate to the variation in the data that is due to the samples under examination. The scores from each data set are then concatenated to form the fused data matrix, followed by autoscaling to properly scale and weight the variables. Unlike with the low-level procedure, the fused matrix is significantly smaller than the original data sets—in most cases, a reduction of about 99%. By using data reduction methods, the relevant data is retained using fewer variables than in the original data set. Weighting and scaling are simply accomplished by use of autoscaling, alleviating some of the complication of low-level methods. Medium-level data fusion enables the fusion of multivariate measurements with univariate measurements, enabling the potential fusion of Raman spectra with temperature, pressure, and flow rates. This is not possible using low-level data fusion. Variable selection methods such as interval PLS (iPLS) [8,9] and genetic algorithms [10] provide alternate means of reducing the initial data to the most pertinent variables. Typically, models are built using the selected variables to determine the success of the selections. After selection, the variables from each data are fused to form a fused spectra, which then undergoes classification or quantification. Once a fused matrix has been formed, additional rounds of variable selection can be performed to determine the importance of the variables or scores selected with regard to the fused matrix. Medium-level data fusion has been successfully employed for a wide array of applications ranging from the deployment of electronic noses for the analysis of beer [11] to the analysis of pigments in works of art [12]. The major disadvantage of medium-level methods is that they require a larger number of models to be generated before fusion. Each data set must be reduced before a final calibration model is produced, and this significantly increases the complexity of the process. The success or failure of the final fused model depends heavily on the models used to reduce the data sets to the scores. Additionally, medium-level data fusion can employ PLS to develop the scores. This means an additional set of analyses is required to generate the necessary reference information, again increasing the model's complexity.

12.1.3   HIGH-LEVEL

High-level data fusion proceeds by preprocessing the individual data sources in the same manner as the low and medium approaches. Following preprocessing, the individual data is used to build models. At this stage, high-level data fusion deviates from

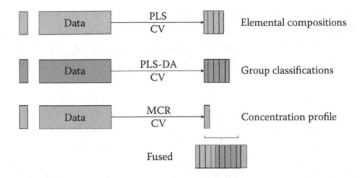

FIGURE 12.7 Schematic demonstrating a high-level data fusion procedure.

previous levels. The modeling techniques applied can vary significantly and be tailored to data-specific needs and goals [13,14]. Figure 12.7 shows examples of the possible modeling scenarios that can be employed. Outputs from each of the models are then fused together to form the final fused matrix. The fused matrix then undergoes analysis to build a final model for control and decision making. Use of high-level data fusion opens the door to a level of fusion that allows fusion of both classification and quantification models. The success of high-level data fusion depends on the models used to generate the data for fusion. Nonoptimal construction or overfitting will lead to nonoptimal and overfit fused models. A propagation of error also occurs, insofar as the errors from each of the initial models will be combined within the fused model. Initial model dependence and propagation of errors have led to a limited application of high-level data fusion.

Two applications will be presented that highlight the feasibility and value that data fusion techniques can provide within a petrochemical context. The first application demonstrates the use of low-level data fusion for the characterization of crude oil fractions. The second application involves the use of medium-level data fusion for the in-depth characterization of whole crude oils.

12.2 APPLICATION 1—CHARACTERIZATION OF CRUDE OIL FRACTIONS USING LOW-LEVEL DATA FUSION

The main goal of this investigation was to determine the feasibility of data fusion for the characterization of crude oil fractions. IR, Raman, and NMR spectroscopies were employed to build data fusion models. IR and Raman techniques can be considered to be complimentary, as both methods examine vibrations of functional groups. NMR spectrometry is a technique orthogonal to IR and Raman, as it provides information relating to the electronic structure of a molecule. Feasibility was evaluated by comparing models built using data fusion approaches, with individual models built from each of the spectroscopic methods. PLS calibration and validation models were produced, and the quality of the final models was assessed with root mean square error of prediction (RMSEP).

$$\text{RMSEP} = \sqrt{\frac{\sum (\text{predicted} - \text{actual})^2}{N-1}} \qquad (12.1)$$

Five different crude oil samples collected from different field locations were supplied by Chevron Energy Technology Company (Chevron ETC). Each crude oil sample was distilled into a number of fractions of sequentially higher boiling point ranges; the three lowest boiling point fractions were analyzed as part of this study. All of the distillates were transferred to 50-mL glass scintillation vials for spectral analysis. Four process parameters were modeled: °API, crude yield by wt.%, crude yield by vol.%, and hydrogen content. The associated reference information for the process parameters was determined using the appropriate American Society for Testing and Materials (ASTM) method. °API is a measure that compares the density of crude oil relative to water. The °API value is calculated from the specific gravity of the crude oil by using an equation that spreads the apparent range of values. Use of °API allows quick comparisons of different crude oils, allowing distinction between lighter (having higher concentrations of shorter-chain hydrocarbons) and heavier (containing higher concentrations of aromatic and condensed hydrocarbons) oils. °API provides a relative measure of the value of crude oils, as lighter crude oils (high APIs) have a greater market value. Therefore, accurate and rapid acquisition of °API for crude oil will enable critical decisions regarding refining and sales of crude oil to occur in a shorter time frame, saving significant resources and maximizing profit. Weight percentage and crude yield by volume percentage are indicators of how much of each fraction is present within the whole raw crude. Rapid determination of the volume of lighter (and greater-value) fractions can significantly influence decisions made in terms of refining and postrefining market value. These parameters are key components in determining the value and final quality of the refined petroleum products; therefore, it was vital to produce models with the highest degree of accuracy and confidence. This required minimizing all modeling errors to the smallest possible values.

All IR spectra were acquired using a ReactIR iC10 FT–IR process spectrometer from Mettler Toledo (Columbus, OH). This was equipped with a silver halide fiber optic diamond attenuated total reflectance immersion probe and a mercury cadmium telluride detector. Each IR spectrum obtained was an accumulation of 1024 scans at a resolution of 2 cm^{-1}, collected from 650 to 2000 cm^{-1}. Use of the immersion probe meant that the crude oil could be sampled directly within the scintillation vial without further need for sample pretreatment. The Raman spectra were collected using a Kaiser Optical Systems RXN1000 process Raman instrument equipped with a 1000-nm laser (Ann Arbor, MI). The instrument was equipped with a Raman Hastelloy immersion ballprobe with UV-grade sapphire optical interface manufactured by Matrix Solutions (Seattle, WA). Each final spectrum was the total of 12 cumulative scans, each with an exposure time of 5 s per scan. The ^{1}H NMR spectra were collected with an offline Bruker AVANCE 500-MHz NMR spectrometer (Billerica, MA) with a total analysis time of approximately 7 min. The crude oil samples were prepared in solutions of 1% tetramethylsilane (TMS) and CDCl$_3$. The high resolution of the data compared with that of the IR meant the NMR had to be mathematically downsampled to allow for fair comparisons and fusion. All of the spectra collected were exported from the respective software packages into MATLAB® 7.5 (The Mathworks Inc., Natick, MA). Once imported, the NMR spectra were downsampled from 500 to 60 MHz using an in-house routine to approximate the spectra obtained from a process NMR instrument. The Raman spectra were corrected for baseline variations and fluorescence using the Polyfit algorithm [15]. The IR spectra were processed using multiplicative scatter correction routines to account for the variations observed in the baseline.

The spectra for each fraction across three boiling point ranges for each spectroscopic method are shown in Figure 12.8. The IR spectra show that as the boiling point increases, there is a reduction in the fingerprint region consistent with the greater saturation and increased homogeneity of the higher boiling fractions. The NMR spectra further demonstrate the increasing quantities of conjugated and aromatic species found in the higher boiling fractions. The NMR spectra for the high boiling fractions show peak groupings found just after 2 ppm and between 6 and 8 ppm that are associated with longer-chain alkanes and aromatic species; these groups are not as evident in the spectra from the lower boiling fractions. The increase in conjugated and aromatic species found in the higher boiling fractions has the effect of increasing the amount of potentially fluorescing material within the samples. This is evident in the Raman spectra. The effect of fluorescence on the Raman spectra was reduced by using a 1000-nm laser source and application of the Polyfit algorithm to maximize the amount of Raman information recovered from the background signal.

The first stage of data fusion required normalization all of the spectra to ensure that each spectrum has the same unit area. Normalization reduces the magnitude

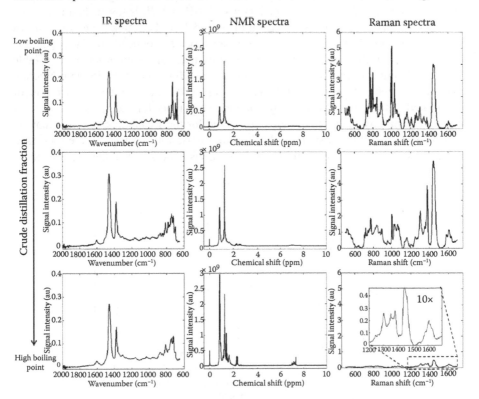

FIGURE 12.8   Spectra demonstrating the differences observed in each fraction for the different spectroscopic methods.

effect of the larger signal intensity of the NMR relative to the vibrational spectra. The result is that each spectrum carries the same statistical weight within a model. A typical fused spectrum is shown in Figure 12.9; the first section of the spectrum is the IR data covering 1399 data points. The second portion of the spectrum is the NMR spectrum and contains 3897 data points. The final portion of the spectrum is the Raman data and accounts for 1799 data points. After concatenation, the fused matrix is mean centered and processed using PLS [16,17]. The modeling with PLS was performed using routines from the PLS Toolbox (Eigenvector Research, Wenatchee, WA). Calibration models were produced using 80% of the data collected. The remaining 20% was used for model performance testing and validation. The final quality of the models was determined using the RMSEP. The selection of samples for calibration and prediction was iterated 500 times to prevent bias or leverage due to the samples selected. Models were produced for each of the spectroscopic methods; one model was built by fusing the Raman and IR data, and another model was built by fusing all three spectroscopic methods together.

The models built using the IR data had an average RMSEP of 3.68% for the four parameters. A representative model for °API built from IR data is shown in Figure 12.10a. The individual errors of each of the quality control parameters are shown in Table 12.1. The model building procedure was repeated using the Raman spectra. The mean RMSEP for these models was 1.81%, which is lower than the errors determined for the IR models. A typical PLS model for the Raman spectra is shown in Figure 12.10b. Overall, the IR and Raman models produced similar prediction errors, except in the case of percentage yield by mass and volumes shown in Table 12.1. Modeling with the Raman spectra produced a 60% reduction in the prediction error associated with percentage yield by mass and a 58% reduction in the error associated with percentage yield by volume. The errors for the remaining parameters were slightly higher than those calculated for the IR models, and this may be attributed to the reduced signal-to-background noise ratio caused by the increased fluorescence of the higher boiling point fractions. The same modeling procedure

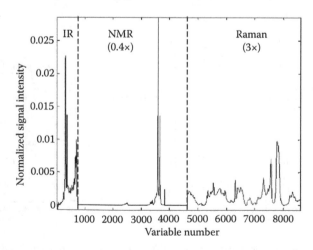

FIGURE 12.9 Typical fused spectrum; signal intensities scaled for clarity.

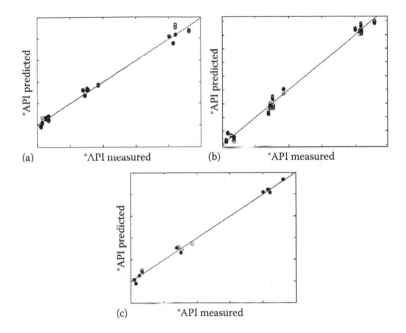

FIGURE 12.10 Typical prediction models for °API built with IR (a), Raman (b), and NMR (c) spectra.

TABLE 12.1

Summary of Prediction Errors from Different Forms of Models Built for Prediction of Process Parameters

| | Average RMSEP (%) | | | | |
Quantity	IR	Raman	NMR	Fused IR and Raman	Fused IR, NMR, Raman
°API	1.845	1.960	0.477	0.381	0.237
% Yield by weight	6.358	2.495	0.844	1.695	0.451
% Yield by volume	6.044	2.089	0.863	1.613	0.510
H$_2$ content	0.463	0.708	0.063	0.136	0.030

was applied to the NMR spectra resulting in an average RMSEP of 0.562% for all four parameters. A typical PLS model built using the NMR data is shown in Figure 12.10c. It is evident that the NMR more effectively models all four of the quality control parameters. The excellent NMR model performance can be attributed to more reproducible offline measurements and the nature of the NMR not being negatively affected by the fluorescent nature of the higher boiling fractions.

The feasibility and effectiveness of data fusion approaches was investigated through the development of two models. The first model of fused data used only the vibrational spectra (Raman and IR) and combined them to produce a complementary fusion model. The second model of fused data used all three forms of spectral data, adding the orthogonal NMR component to the vibrational models. The PLS models were built using the same method as was performed with the individual spectroscopic models. The average RMSEP calculated for the fused IR and Raman models was 0.956%. An example of the fused vibrational spectroscopy model is shown in Figure 12.11. The modeling error of the fused vibrational data was significantly reduced when compared with either of the individual IR or Raman models. This decrease in model error can be attributed to the fused model capturing the variance and correlations that occur between the individual models. The fused IR and Raman °API model prediction errors were reduced by an astonishing 79%. The data presented in Figure 12.11 also demonstrates that the fusion of Raman and IR spectra more efficiently models the higher end of the °API scale than the individual models. By fusing the individual spectroscopic measurements together, the negative impact that the higher boiling point fractions had on the IR model was greatly reduced by the addition of the Raman spectra. The variation associated with each individual spectrum was minimized by fusing them together. The second fused model, which included the vibrational spectra and NMR data, demonstrated an additional increase in performance and reduction of prediction error when compared with either of the individual models and the IR/Raman fused model. The average %RMSEP for the full fused model was calculated to be 0.307% (Table 12.1) and is shown in Figure 12.12. By adding the NMR data to

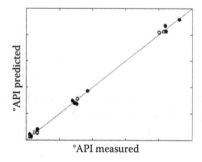

FIGURE 12.11 Typical model for the prediction of °API using a fusion of Raman and IR data.

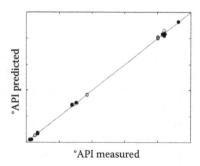

FIGURE 12.12 Example PLS model for prediction of °API from a fusion of Raman, IR, and NMR spectra.

the Raman and IR data, all samples are predicted to have a high degree of accuracy and very low RMSEP when compared with the previous models. The orthogonality of NMR decreased the modeling error by approximately a factor of 3 when compared with the IR and Raman fused model. The orthogonality of the NMR data greatly enhanced the amount of chemical information captured in the model. This is apparent with the reduction in error seen in the models used to predict the hydrogen content of the fractions. Proton NMR spectra represent the presence of hydrogen atoms and their spatial relationships. By including NMR data within the model, the chemical information describing the hydrogen content was effectively increased. The augmentation of the information describing the hydrogen content resulted in a 78% reduction in the RMSEP when compared with the fused vibrational models. The significant reduction of error relates to a significant increase in the value of the spectroscopic methods and the modeling procedures for online applications. The errors in prediction for the fused vibrational and NMR models are highlighted in bold Table 12.1.

The effective fusion and modeling of Raman, IR, and NMR data succeeded in greatly improving the model performance and decreasing the modeling error for the prediction of key quality parameters used for controlling the refining of crude oil. Furthermore, the addition of the orthogonal NMR data significantly improved the models and added value to the measurements performed. The application of data fusion gives a user an effective means of making a large number of meaningful measurements, and then extracting all relevant information and variation from the acquired data quickly and efficiently. The use of data fusion with multiplexed analytical measurements clearly reduces the error associated with inferential property models for streams with highly complex compositions, such as crude oil, its fractions, and petrochemical products.

12.3 APPLICATION 2—CHARACTERIZATION OF WHOLE CRUDE OIL USING MEDIUM-LEVEL DATA FUSION

The primary goal for this application was to build on the foundation of work performed in application 1 by progressing the data fusion procedures to model whole crude samples across a wider array of process parameters. The analysis of whole crudes presented a greater level of complexity than seen in application 1, as the samples provided varied significantly in terms of color, viscosity, and density. The feasibility of the different spectroscopic methods employed in application 1 were reexamined as a greater amount of aromatics, condensed hydrocarbons, and long-chain aliphatic species are present in whole crude samples. Data reduction through the use of variable selection algorithms meant that this application would be employing medium-level data fusion techniques. Model-to-model comparison for each of the parameters was performed using the relative standard error in prediction (RSEP) due to the differing magnitudes and variances of the process parameter under investigation.

$$\text{RSEP} = \sqrt{\frac{\sum (\text{predicted} - \text{actual})^2}{\sum \text{actual}^2}} \times 100 \qquad (12.2)$$

Fifty whole crude oil samples, collected from different locations worldwide, were provided by Chevron ETC for the purposes of model building and calibration. A further 10 samples were provided for validation, some of which were duplicates from within the calibration set. All samples were provided in 50-mL scintillation vials for analysis with immersion probes. Eleven process parameters were investigated, which included hydrogen, nitrogen, sulfur, °API, MCR, and viscosity. MCR is a measure of the relative capability of the crude oil to deposit carbon residue. The ability to predict MCR rapidly and accurately can have significant impact on the refining process. Accurate prediction of sulfur content has value to the refining process because government regulations regarding sulfur content of refined fuels are extremely restrictive. It has also been shown that sulfur dioxide can potentially poison the metal catalysts used during the catalytic reforming process. The associated reference information was collected using the appropriate ASTM methods.

All of the IR spectra were collected using a ReactIR iC10 FT–IR process spectrometer from Mettler Toledo (Columbus, OH), equipped with a K6 conduit immersion probe. Each IR spectrum obtained was an accumulation of 1024 scans at a resolution of 2 cm^{-1} collected from 650 to 4000 cm^{-1}. Raman spectra were collected on two instruments: a Kaiser RXN2 equipped with a 785 nm excitation source and a Kaiser RXN1000 equipped with a 1000 nm excitation source. Both instruments were equipped with a Raman Hastelloy immersion ballprobe with a UV-grade sapphire optical interface manufactured by MarqMetrix (Seattle, WA). Each Raman spectrum was the accumulation of 12 scans with exposures of 5 s per scan. The ^{1}H NMR spectra were collected with an offline Bruker AVANCE 500 MHz NMR spectrometer with a total analysis time of approximately 7 min. The crude oil samples were prepared in solutions of 1% TMS and CDCl$_3$. As previously outlined in application 1, the NMR spectra were mathematically reduced to approximate the process instrument resolution. All of the spectra collected were exported from the respective software packages into MATLAB® 7.5 (The Mathworks Inc.).

Figure 12.13a and b show the Raman spectra collected of the whole crudes. The increased concentration of aromatic species present in the whole crudes, when compared with the fractions, caused a massive increase in the background fluorescence. The increased fluorescence signal completely eradicated the Raman signals, making Raman spectroscopy infeasible for the analysis of the whole crude samples and was therefore removed from further analyses. Figure 12.14 shows the IR spectra collected of the calibration and validation samples. Peaks relating to C–H stretches can be observed at 2971, 1604, 1442, and 1377 cm^{-1}. A number of stretches relating to C–H, C–C, and C=C can be found from 1000 to 650 cm^{-1}. The NMR spectra (Figure 12.15) contained significant contributions and variation from aliphatic hydrogens (0–3 ppm) and aromatic hydrogens (6–8 ppm).

The medium-level data fusion began with the processing of the IR and NMR datasets. Extended multiplicative scatter correction [18–20] and mean centering were applied to IR spectra to remove unwanted variation. The NMR spectra first underwent correlation optimized warping alignment using LineUp (Infometrix Inc., Bothell, WA) followed by mean centering to removed instrumental variability. A number of variable selection routines were examined as methods of data reduction. Techniques such as iPLS, genetic algorithms, and D-optimality were examined in terms of time taken for

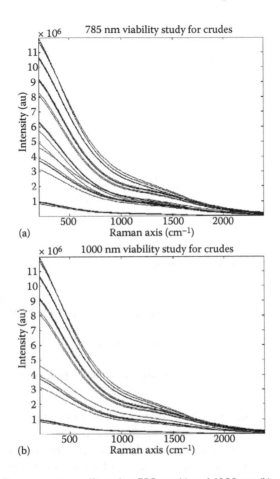

FIGURE 12.13   Raman spectra collected at 785 nm (a) and 1000 nm (b).

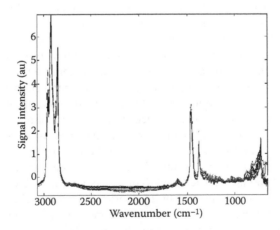

FIGURE 12.14   IR spectra of whole crudes.

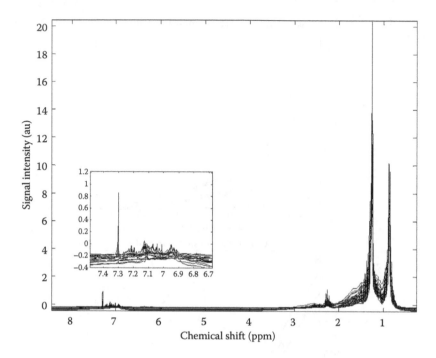

FIGURE 12.15   NMR spectra of whole crudes. Aromatic region 6–8 ppm (inset).

selections and final model prediction errors. iPLS was found to be the optimal method for variable selection. iPLS performs variable selection by two main paths: forward selection or backward elimination. In the forward selection method, the spectra are split into subsections and PLS models are built for each subsection. The subsection with the smallest error forms the first set of variables selected. Following this, each individual subsection is added to the first set and a PLS model is built for the pair. The pair that has the smallest prediction error forms the second set of selected variables. This procedure of concatenation, modeling, and selection continues until a predefined number of subsections has been selected or there is no improvement in modeling error of the selected subsections. Backward elimination methods proceed by splitting the spectra into a number of subsections; each section is removed, in turn, and a PLS model is built for the remaining subsections. After each subsection has been removed and modeled, the section that after removal improved the model the most is eliminated from the variable set. Once elimination occurs, the process of removing subsections, modeling, and elimination occurs until a predefined number of subsections have been removed or the model errors of the selected samples stop improving.

After variable selection, the subsets of variables of each data set are concatenated to form a fused data matrix. The fused data matrix and the reference data from the 11 process parameters were used to build a series of PLS models. The fused matrix and reference data were mean centered before modeling, and the latent variables (LVs) used with each were optimized through cross validation. The RSEP for each of the 11 models is shown in Figure 12.16. The chemical properties (hydrogen content,

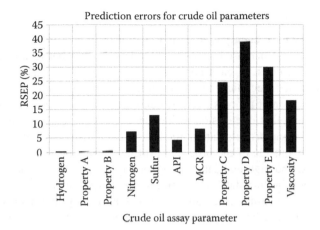

FIGURE 12.16   Chart of RSEP for 11 crude oil parameters.

properties A and B, and nitrogen) are predicted accurately in the respective models. The success of the modeling can be attributed to the amount of information present in the original data sets and the subsequent performance of the iPLS to maximize the relevant information within the fused matrix. Conversely, the model produced for sulfur content had relatively poor predictive performance relative to the other chemical properties. This is due, in part, to the lack of information within the original data sets pertaining to sulfur content; neither IR nor proton NMR contains the region of sulfur sensitivity. Most of the ASTMs generated for sulfur testing require analysis to be performed using x-ray, electrochemical, or thermal-based procedures. Prediction of sulfur through data fusion will improve significantly by including x-ray fluorescence or atomic spectroscopy data within the fused matrix, allowing effective capture of information pertaining to sulfur. The models predicting °API and MCR performed well, both models with RSEP of <10%. The °API models were accurate to within <5% of the °API; this equates to ±1.8°API. Physical properties (properties C, D, and E, and viscosity) were predicted with less success. A portion of the high prediction error can be linked with the relatively higher errors found in the reference data for these physical properties when compared with the chemical properties. Furthermore, each of the physical properties exhibits nonlinear qualities; improvement of the predictive abilities of the models will be achieved through the use of locally weighted regression or poly-PLS to better capture nonlinear characteristics.

Building on the successful foundations laid out in application 1, data fusion was used to characterize a series of crude oil samples. Models of chemical and physical properties were produced that had a high degree of accuracy, in particular models pertaining to the prediction of chemical properties A, B, and hydrogen. In terms of crude oil characterization, the data fusion procedure can be extended to include GC, x-ray fluorescence, and laser-induced breakdown spectroscopy. The application of data fusion may also make it possible to produce accurate predictions of the properties of crude oil fractions based on the spectra collected of the whole crude oils, especially by expanding the fused data to include a wider variety of information sources.

12.4 SUMMARY

Two applications have been presented that demonstrate the enhancements in modeling and, therefore, increased control and understanding that results from the use of data fusion methods. In each application, the use of data fusion outperformed traditional "unfused" techniques in terms of modeling performance and applicability. The proliferation of process analyzers and the increased rate at which data can be collected mean that data fusion techniques will probably become the standard method for analyzing process data in the future. Data fusion will also begin to gain traction in fields outside of the process world, areas such as website optimization, sports analytics, and high-end data mining, or any arena where exploiting the information contained in "big data" is becoming the modus operandi. The techniques described in the above-described applications require only that data be consistent, contain variation, and be independent of the application to which they are being applied.

ACKNOWLEDGMENTS

This work was supported by Chevron ETC and the Applied Physics Laboratories (University of Washington, Seattle). The authors would like to thank Kaiser Optical Systems and Mettler Toledo for their support and supply of instrumentation. The authors would also like to thank Dr. Ajit Pradhan and Brian Morlan of Chevron ETC for providing the NMR and crude assay data.

REFERENCES

1. Workman, J., Jr. 1993. A review of process near infrared spectroscopy 1980–1994. *Journal of Near Infrared Spectroscopy* 1 (4):221–245.
2. Workman, J., Jr. 1996. Review: A brief review of near infrared in petroleum product analysis. *Journal of Near Infrared Spectroscopy* 4 (1):69–74.
3. Nordon, A., C. A. McGill, and D. Littlejohn. 2001. Process NMR spectrometry. *Analyst* 126 (2):260–272.
4. Nordon, A., C. A. McGill, and D. Littlejohn. 2002. Evaluation of low-field nuclear magnetic resonance spectrometry for at-line process analysis. *Applied Spectroscopy* 56 (1):75–82.
5. Andrade, J. M., S. Muniategui, and D. Prada. 1997. Prediction of clean octane numbers of catalytic reformed naphthas using FT–m.i.r. and PLS. *Fuel* 76 (11):1035–1042.
6. Cooper, J. B., K. L. Wise, W. T. Welch, R. R. Bledsoe, and M. B. Sumner. 1996. Determination of weight percent oxygen in commercial gasoline: A comparison between FT–Raman, FT–IR, and dispersive near-IR spectroscopies. *Applied Spectroscopy* 50 (7):917–921.
7. Bueno, A., C. A. Baldrich, and D. Molina. 2009. Characterization of catalytic reforming streams by NIR spectroscopy. *Energy & Fuels* 23:3172–3177.
8. Norgaard, L., A. Saudland, J. Wagner, J. P. Nielsen, L. Munck, and S. B. Engelsen. 2000. Interval partial least-squares regression (iPLS): A comparative chemometric study with an example from near-infrared spectroscopy. *Applied Spectroscopy* 54 (3):413–419.
9. Leardi, R., and L. Norgaard. 2004. Sequential application of backward interval partial least squares and genetic of relevant spectral regions. *Journal of Chemometrics* 18 (11):486–497.

10. Chapter 27 Genetic algorithms and other global search strategies. 1998. In *Data Handling in Science and Technology*, edited by D. L. Massart, B. G. M. Vandeginste, L. M. C. Buydens, S. De Jong, P. J. Lewi and J. Smeyers-Verbeke. Elsevier.

11. Vera, L., L. Acena, J. Guasch, R. Boque, M. Mestres, and O. Busto. 2011. Discrimination and sensory description of beers through data fusion. *Talanta* 87:136–142.

12. Ramos, P. M., and I. Ruisanchez. 2006. Data fusion and dual-domain classification analysis of pigments studied in works of art. *Analytica Chimica Acta* 558 (1–2):274–282.

13. Doeswijk, T. G., A. K. Smilde, J. A. Hageman, J. A. Westerhuis, and F. A. van Eeuwijk. 2011. On the increase of predictive performance with high-level data fusion. *Analytica Chimica Acta* 705 (1–2):41–47.

14. Silvestri, M., L. Bertacchini, C. Durante, A. Marchetti, E. Salvatore, and M. Cocchi. 2013. Application of data fusion techniques to direct geographical traceability indicators. *Analytica Chimica Acta* 769:1–9.

15. Afseth, N. K., V. H. Segtnan, B. J. Marquardt, and J. P. Wold. 2005. Raman and near-infrared spectroscopy for quantification of fat composition in a complex food model system. *Applied Spectroscopy* 59 (11):1324–1332.

16. Wold, S., H. Martens, and H. Wold. 1983. The multivariate calibration-problem in chemistry solved by the PLS method. *Lecture Notes in Mathematics* 973:286–293.

17. Kowalski, B. R., and M. B. Seasholtz. 1991. Recent developments in multivariate calibration. *Journal of Chemometrics* 5 (3):129–145.

18. Decker, M., P. V. Nielsen, and H. Martens. 2005. Near-infrared spectra of *Penicillium camemberti* strains separated by extended multiplicative signal correction improved prediction of physical and chemical variations. *Applied Spectroscopy* 59 (1):56–68.

19. Kohler, A., C. Kirschner, A. Oust, and H. Martens. 2005. Extended multiplicative signal correction as a tool for separation and characterization of physical and chemical information in Fourier transform infrared microscopy images of cryo-sections of beef loin. *Applied Spectroscopy* 59 (6):707–716.

20. Martens, H., and E. Stark. 1991. Extended multiplicative signal correction and spectral interference subtraction—New preprocessing methods for near-infrared spectroscopy. *Journal of Pharmaceutical and Biomedical Analysis* 9 (8):625–635.

13 Application of Computer Simulations to Surfactant Chemical Enhanced Oil Recovery

*Jan-Willem Handgraaf, Kunj Tandon, Shekhar Jain,
Marten Buijse, and Johannes G.E.M. Fraaije*

CONTENTS

CONTEXT

Enhanced chemical oil recovery (EOR) allows increased extraction of oil as oil wells follow their normal decline curves. Determining the best mixture of chemicals for EOR can be aided by simulations that

- Predict the properties of oil–brine–surfactant emulsions
- Allow fast screening of the possible EOR chemical mixtures

ABSTRACT

Surfactant formulations are currently being used by the oil industry in chemical enhanced oil recovery. Given a properly selected formulation, a microemulsion is formed between the (crude) oil and brine that resides in the reservoir. The formed microemulsion typically has an ultralow interfacial tension (~10^{-3} mN/m) that will restore the oil flow in the reservoir. Here we discuss a simulation protocol for computing the properties of oil–brine–surfactant microemulsions. The protocol is based on dissipative particle dynamics, a coarse-grained simulation technique that allows for a fast screening of parameter space. The "Method of Moments" [Fraaije et al., *Langmuir* 2013, 29, 2136–2151], which looks at the bending properties of the surface film, is introduced and applied to a set of anionic sulfonate and sulfate surfactants. For the set of surfactants, we compute the optimum salinities and address the effect of surfactant tail length and the co-solvent. We find that the computed optimum salinities for neat oil–brine–surfactant interfaces quantitatively agree with experimentally obtained values. In addition, the Method of Moments correctly predicts that the optimum salinity decreases with increasing surfactant tail length. Furthermore, the simulations show that the addition of a co-solvent leads to lower salinity, again in good agreement with experiments. We finally discuss how to extend the Method of Moments to industrially relevant mixtures that include surfactant mixtures and crude oils.

13.1 INTRODUCTION

Chemical enhanced oil recovery (cEOR) is a method of displacing the remaining capillary-trapped oil from an oil reservoir to the production well after regular methods, such as water flooding, are depleted. The surfactant mixture is usually injected in the field as a dilute solution in brine, often in combination with other chemicals such as alkali, polymers, and co-solvents to further improve the oil displacement to the production well. In the past, EOR methods and related experimental and simulation studies of microemulsions have seen several revivals (see, e.g., References 1–3). In all cases, these revivals are related to peaks in the crude oil price. For a recent overview of the application of EOR in the oil industry, we refer to the work of Hirasaki et al. [4].

Given a properly selected surfactant–polymer–co-solvent combination at optimum salinity, temperature, and pressure, a microemulsion (or middle) phase will form as a third phase in the crude oil–brine–surfactant system (Figure 13.1).

The main idea of surfactant cEOR is to have an as large as possible microemulsion phase, thereby optimizing the oil yield at the production well. The interfacial tension (IFT) in this three-phase oil–brine–surfactant system can be as low as 10^{-3} mN/m, roughly four magnitudes lower than an oil–brine system in the absence of surfactants (~50 mN/m). To complicate matters, the crude oil type will determine the efficacy of a surfactant mixture and the resulting IFT of the system, and hence must be specially tailored to the crude oil composition and its secondary properties such as salinity, alkalinity, temperature, and pressure. The surfactant screening process is usually laborious, and many experiments such as phase behavior or IFT spinning drop tests must be performed. Although extensive guidelines have been developed during the last 50 years, mostly obtained through trial and error, surfactant selection is still very much a time-consuming and costly process. Furthermore, detailed understanding of the complex molecular interactions at the interface between oil, surfactant, polymer, co-solvent, and brine is usually completely lacking. For a recent review of surfactant EOR, we refer to the work of Salager et al. [5].

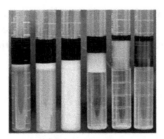

FIGURE 13.1 Example of industrial phase behavior scan of a twin-tailed sulfonate surfactant with a crude oil. Surfactant concentration: 0.5 wt.%. Tests with 4 mL crude and 10 mL surfactant solution at six different NaCl salinities, from left to right: 1.0, 1.2, 1.4, 1.6, 1.8, and 2.0 wt.%. Optimum salinity is at 1.6% NaCl with a middle-phase microemulsion in equilibrium with the oil phase and brine phase. Surfactant is a commercial internal olefin sulfonate (IOS), with an alkyl tail length in the range 20–24, manufactured by Shell Chemicals. The crude oil is a light oil (density = 0.82 g/cm^3, viscosity = 1.2 mPa·s) from a South East Asian reservoir. The oil composition is as follows: 59.2% saturates, 33% aromatics, 7.6% resins, and 0.2% asphaltenes. The test temperature is 69°C. (From Shell Laboratories.)

In this chapter, we will first briefly outline the two simulation techniques required to study microemulsions by computational means. Both techniques have their basis in the 1990s. On the basis of these two simulation techniques, we will introduce a new method called the "Method of Moments" (MoM), which allows for a fast and accurate screening of oil–brine–surfactant systems. We then, very briefly, give the computational details, followed by the results with a focus on the role of the surfactant and the co-solvent. We end the chapter with some conclusions and an outlook on the applicability of molecular simulation techniques to surfactant cEOR.

13.2 DISSIPATIVE PARTICLE DYNAMICS

Dissipative particle dynamics (DPD) was developed by Shell researchers Hoogerbrugge and Koelman in the early 1990s [6], and subsequently fine-tuned by Español and Warren [7] and Groot and Warren, working for Unilever [8]. The DPD method has been applied to a huge variety of molecular systems, ranging from simple liquids to (very) large colloidal systems and vesicles. It is outside the scope of this work to discuss the application of DPD. Here, we just name two studies that used DPD to study oil–brine–surfactant interfaces. Using DPD, Rekvig et al. [9] showed that their surfactant model could capture, quite well, the experimental trends observed for single- and double-tailed anionic surfactants. In another study, DPD was used to study interfaces where the surfactant was a linear alkyl ethoxylate. The authors showed that they could get reasonable agreement between simulated and measured IFTs [10].

In DPD, instead of simulating all atomic degrees of freedom, one lumps groups of atoms in so-called coarse-grained particles or beads connected by harmonic springs [11,12]. The coarse graining of the structure and the soft interactions used in DPD simulations allow for larger systems (~1 μm) to be modeled over significantly longer times (~1 ms) than is possible with conventional particle-based simulations methods such as Monte Carlo or molecular dynamics.

In DPD, one solves Newton's equation of motion using either a velocity Verlet integration scheme or more sophisticated integration schemes [13]. The change in velocity of each bead, at every iteration of the simulation, is calculated using the following equation:

$$\partial \mathbf{v}_k = \frac{\mathbf{F}_k^c}{m_k}\partial t + \frac{\mathbf{F}_k^f}{m_k}\partial t + \mathbf{v}_k^r \tag{13.1}$$

where $\partial \mathbf{v}_k$ is the change in velocity of bead k during one time step of the DPD simulation; $\mathbf{F}_k^c$ is the conservative force between bead k and other beads in the model; $\mathbf{F}_k^f$ is the friction force in the system, m_k is the mass of bead k; and $\mathbf{v}_k^r$ is a random velocity that represents thermal fluctuations in the system. It is the friction and random force that distinguish DPD from Brownian dynamics. Both forces are pairwise interactions in DPD, which conserves momentum, and includes hydrodynamics in the simulation.

The conservative force contains all the chemistry of the system and is a sum of three terms: a bead–bead repulsive force, a spring force between connected beads, and an electrostatic force. The friction, or drag force, is defined in DPD as

$$\mathbf{F}_k^f = -\gamma_{kl} m_k \sum_l \mathbf{\Omega}_{kl} \mathbf{v}_{kl} \tag{13.2}$$

where γ_{kl} is the strength of the dissipation between the beads, also known as the friction coefficient, and $\mathbf{\Omega}_{kl}$ is an element of the friction matrix. The friction force is a function of the difference in velocity between bead k and the surrounding beads. A bead that is moving more slowly than those that surround it will be speeded up, whereas one that is moving more quickly will be slowed down.

The pairwise random velocity in DPD is the noise term defined as

$$\mathbf{v}_k^r = \left(\frac{2\gamma_{kl}kT\partial t}{m_k}\right)^{\frac{1}{2}} \sum_l \omega_{kl}\xi_{kl}\hat{\mathbf{r}}_{kl} \tag{13.3}$$

where ω_{kl} is a weighting function and ξ_{kl} is the noise parameter. For all details of the DPD method and parameterization, we refer to the *Culgi Scientific Manual* [14].

13.3 METHOD OF MOMENTS

The MoM, as outlined here, has its origin in the microemulsion theory originally developed by De Gennes and Taupin [15] and Helfrich [16] in the 1980s and 1990s, and later extended by others [17–20]. We define the relation between surface mechanical coefficients (or moments) and the stress profile* as [21–23]

---

* In the following, we assume that the oil–brine–surfactant interface is along the z-dimension of the simulation box.

$$\gamma = M_0 = \int \sigma(z)\,\mathrm{d}z$$

$$\tau = -M_1 = -\int (z - z_0)\cdot\sigma(z)\,\mathrm{d}z \quad (\gamma = 0) \tag{13.4}$$

$$\kappa_s = M_2 = \int (z - z_0)^2 \cdot\sigma(z)\,\mathrm{d}z \quad (\gamma = 0,\ \tau = 0)$$

with the stress profile defined as

$$\sigma(z) = P_\perp - P_\| \tag{13.5}$$

Here γ is the microscopic IFT in N/m, τ the surface torque density in N, and κ_s the splay bending constant in Nm. $P_\perp$ and $P_\|$ are the normal and the tangential pressure components, respectively. In practice, one obtains γ directly from the interfacial brine–oil–surfactant DPD simulations using

$$\gamma = l_z \cdot \left[\langle P_{zz} \rangle - \frac{1}{2}\left(\langle P_{xx} \rangle + \langle P_{yy} \rangle \right) \right] \tag{13.6}$$

where l_z is the box edge in the z-dimension, and P_{xx}, P_{yy}, P_{zz} are the respective diagonal components of the pressure tensor. The angle brackets indicate a time average over the length of the DPD simulation. Similar formulas can be written for the first and second moment. We stress here that the relation for τ given in Equation 13.4 only holds if the IFT is zero. Similarly, the relation for the splay constant only holds if both the IFT and the surface torque density are zero. Alternatively, the surface torque density is defined as [23]

$$\tau \equiv \kappa c_0 \tag{13.7}$$

where κ is the bending constant of the surface film and c_0 is the spontaneous curvature. This means that at optimum conditions, the oil–brine–surfactant interface has no tendency to either bend to the oil or brine phase. From the splay constant, one can deduce the stiffness of the interface, which will be to a large extent determined by the surfactant structure.

In a homogeneous fluid, the stresses of the system average out to zero in all three spatial dimensions. In a heterogeneous oil–brine system, the stress at the interface will be, in general, large (positive), leading to an overall positive IFT (~50 mN/m). When surfactants adsorb at the interface, the shape of the stress profile changes substantially. This is illustrated in Figure 13.2, where $\sigma(z)$ is plotted for an n-alkane–brine interface that is saturated by a twin-tailed sulfonate surfactant as obtained from a DPD simulation.

As discussed before, the oil–brine repulsion leads to a positive contribution to the stress. However, the stress profile now has two negative contributions associated with the favorable interactions between oil and surfactant tails and sulfonate head

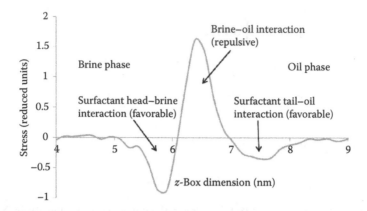

FIGURE 13.2 Stress profile of a twin-tailed sulfonate surfactant at the oil–brine interface obtained from a DPD simulation. $\gamma = 0$, oil is *n*-nonane (C_9). (Adapted from M. Buijse et al., Surfactant optimization for EOR using advanced chemical computational methods, SPE Improved Oil Recovery Symposium, April 14–18, 2012, Tulsa, OK, 154212-MS.)

groups and the brine. The exact shape of the stress profile depends on the (crude) oil type, surfactant characteristics such as tail length, branching, surfactant head type, co-solvent type, the salinity of the brine, and the temperature and pressure in the reservoir. In case the attractive contributions in $\sigma(z)$ completely cancel the repulsive part, a near-tensionless interface is obtained and the IFT is zero.

In the following, we will demonstrate that the DPD coarse-grained simulation technique in combination with the MoM leads to a robust and quantitative computational tool to study the formation of oil–brine–surfactant microemulsions. Note that the MoM is not restricted to DPD, but can be applied to any particle-based simulation technique where the time evolution of the stresses in the system are available. Qualitatively, MoM can be seen as a microscopic double-film model, with the concentration of surfactants at the interface, the oil and tail molecular volumes, and the salt strength of the brine as variables. For all the details of the used simulation method, including a comparison with existing formulation approaches such as the Winsor R theory, quantitative structure property relations, hydrophilic–lipophilic deviation (HLD), and HLD–net average curvature, and temperature coefficients, we refer here to our recently published work [23].

13.4 COMPUTATIONAL DETAILS

A typical DPD simulation is performed in a rectangular simulation box with edges 5 nm × 5 nm × 12 nm, with ~100 anionic coarse-grained surfactant molecules (see Table 13.1 for the molecular structures), 1000 coarse-grained oil molecules in the series *n*-hexane to *n*-pentadecane, and sodium and chloride ions up to the specified salinity. The rest of the simulation box is filled up with water beads (Figure 13.3). The final box density is 3.0 in reduced units [8], and the temperature is set to ambient conditions. We stress here that the aim of this work is not to compute the "macroscopic" IFT of the complete microemulsion, which would consist of several

TABLE 13.1

Molecular Structures and Chemical Formulas of Surfactants Used in This Work

Surfactant	Chemical Formula[a]	Atomic Model	Mesoscopic Model
AOT			
SDS	$CH_3(CH_2)_{11}OSO_3Na$		
n7DTS	$[CH_3(CH_2)_6]_2CHSO_3Na$		
n10DTS	$[CH_3(CH_2)_9]_2CHSO_3Na$		
n13DTS	$[CH_3(CH_2)_{12}]_2CHSO_3Na$		

Source: J.G.E.M. Fraaije et al., *Langmuir*, 29, 2136, 2013.

[a] Chemical formula of AOT taken from Wikipedia.

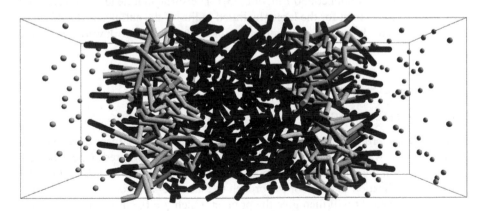

FIGURE 13.3 Snapshot of a simulation box containing a nonane–sodium dodecyl sulfate–brine interfacial system. Particle color coding, black: sulphate head groups, white: surfactant tails, dark gray: oil molecules, light gray: counter ions. Water molecules are not shown. As we are using periodic boundary conditions, the simulation box contains two interfaces. The "packed visualization mode" leads to bonds that will be cut and the edges of the simulation box.

TABLE 13.2

Conversion Factors from DPD to SI Units

M_0 (γ): multiply with 9.86 to get the value in mN/m

M_1 (τ): multiply with 6.36e^{-12} to get the value in N

M_2 (κ_s): multiply with $k_B T = 4.11$e^{-21} ($T = 298$ K) to get the value in Nm

layers of intertwined oil and brine microdomains with surfactant molecules at the interface and in solution. Note that, in principle, such a complex system could be simulated with DPD since simulations with millions of particles are in reach with the current computational power. The simulations presented in this work could be run on a standard desktop computer without problems. Periodic boundary conditions were employed and the smoothed particle mesh Ewald method from Essmann et al. [24] is used to calculate the charge interactions and the ensuing contribution to the stress. The reported IFT results, denoted with the γ symbol (not be confused with γ_{kl}; see Equation 13.2), are in DPD units; see Table 13.2 for conversion factors from DPD units to SI units.

For the coarse graining, we used three heavy atoms per bead. The mapping factor of three is based on our experience with coarse-graining molecular systems [12,25,26]. For example, this means that *n*-nonane is modeled by three beads connected by harmonic strings. Water is modeled by a single bead that corresponds to three water molecules. Hydrated sodium and chloride ions were modeled each by a single bead with corresponding charge. The coarse-grained structures of the surfactant molecules can be found in Table 13.1. The DPD a_{ij} repulsion parameters were kept a-selective ($a = 25$), with the exception of the oil–water repulsion that was set to 80; this value is based on fit to experimental IFT of pure nonane and water (experiment: 52.4 mN/m, simulated: 50.7 mN/m [23]). The surfactant head–head repulsion was set to 35 to account for the larger hydrated radius of the sulfates and sulfonates compared with the bare ones. All simulations presented in this work were performed with the Culgi software package [27].

13.5 RESULTS

13.5.1 Surfactant Selection

The simulation protocol is as follows: surfactant molecules are inserted at the oil–brine interface while the change in IFT is continuously monitored. Once the IFT is close to zero, the simulation is stopped and we extrapolate to $\gamma = 0$. Figure 13.4 shows the application of the MoM approach to a nonane–aerosol OT (AOT)–brine interface. A linear fit will then give the point at which $\gamma = 0$. Clearly this approach has the advantage that there is no need to actually perform a simulation where γ ~10^{-3} Nm/m, which is inherently difficult to perform with particle-based simulation techniques. We can simply shoot through zero and still extrapolate to zero. Once the condition $\gamma = 0$ is satisfied, we start to change the salinity by introducing sodium and chloride ions in equal amounts to the brine phase while the first moment of the stress,

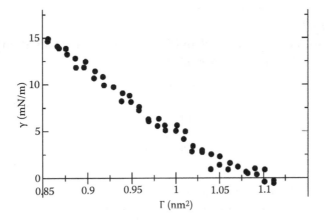

FIGURE 13.4 Zeroth stress moment, M_0 (γ), of a nonane–AOT–brine interface as a function of the surface area per surfactant (Γ). Salt strength is that for a balanced microemulsion with a salinity of $\log_{10} C(M) = -1.625$. (Adapted from J.G.E.M. Fraaije et al., *Langmuir*, 29, 2136, 2013.)

the so-called surface torque density (τ), is now monitored. The protocol outlined above is very robust and does not require having a completely flat interface during the simulation, which in fact is not the case, even for these relatively small simulation box dimensions [23].

In Figure 13.5, we have plotted the surface torque density (τ) as a function of the salinity for the set of anionic surfactants (see Table 13.1, for the molecular structures), and in Table 13.3 we have summarized the optimum salinity results.

Overall, we obtain a very good agreement with experiments. SDS has a positive surface torque for the complete salinity range and, hence, will not form a microemulsion.

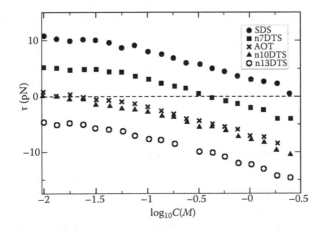

FIGURE 13.5 Calculated surface torque density (τ) of tensionless nonane–surfactant–brine interfaces as function of brine salt concentration. (Adapted from J.G.E.M. Fraaije et al., *Langmuir*, 29, 2136, 2013.)

TABLE 13.3

Comparison of Experimental and Simulated Optimal Salt Strengths ($\log_{10} C(M)$) for the Set of Anionic Surfactants

Surfactant	Experimental	Predicted
SDS	ND	ND
n7DTS	−0.4 ... −0.5[a]	−0.3
AOT	−1.8[b]	−1.7 ... −1.8
n10DTS	−1.8 ... −2.0[a]	−1.9
n13DTS	ND	ND

Source: J.G.E.M. Fraaije et al., *Langmuir*, 29, 2136, 2013.

Note: See Table 13.1 for the molecular structures. ND, not detected.

[a] Value taken from R. Granet et al., *Colloids Surf.*, 49, 199, 1990.

[b] Value taken from R. Aveyard et al., *J. Chem. Soc. Faraday Trans. I*, 82, 125, 1986; and O. Ghosh and C.A. Miller, *J. Phys. Chem.*, 91, 4528, 1987.

This fact is well known as SDS needs a co-solvent such as an alcohol to form a microemulsion [29,30]. For the double-tail sulfonates (DTS), there is experimental data for n7DTS and n9DTS with $\log_{10} C(M) = −0.4$ ... $−0.5$ and $−1.1$ ... $−1.2$, respectively [28]. The values we compute for n7DTS and n10DTS are in good agreement, if we extrapolate the value for n9DTS in the latter case. The experiments show further that n11DTS does not have an optimum. Similarly, we do not find an optimum for n13DTS that has, for the complete salt range, a negative torque, indicative of a strong tendency of the surfactant to bend toward the brine phase. We stress here that the DTS surfactants used in this work and in Reference 28 are pure components that differ substantially from what one currently uses in field tests. Typically, one would use a mixture of surfactants with a variation not only in tail length but also in the position of the sulfonate head group along the alkyl chain. In addition, a small fraction of the surfactants in the mixtures will be disulfonates that would be better soluble in brine, and hence, would lead to higher optimum salinities. For example, Barnes et al. [31] showed that for the so-called ENORDET surfactant mixtures, as developed by Shell, the optimum salinities are considerably higher. For the series IOS C15–C18, IOS C20–24, and IOS C24–C28, they report $\log_{10} C(M) = 0.4$ ... 0.2, $−0.2$ ... $−0.5$, and $−0.5$ ... $−0.8$, respectively.

13.5.2 EFFECT OF SURFACTANT TAIL LENGTH

If we look at the series n7DTS, n10DTS, n13DTS, we find that the average surface area per surfactant (a_s) increases from 0.44 to 0.57 nm². This is to be expected, as the packing of the surfactant molecules at the interface will decrease as a function of the tail length. This can be rationalized using the so-called molecular packing parameter that has its origin in the theoretical and experimental study of self-assembly of surfactant molecules at interfaces. In Reference 32, Nagarajan argues that not only the surfactant head group but also the surfactant tail influences the surface area per surfactant, and ultimately the shape of the aggregates in solution. Interestingly, molecular dynamics simulations showed that for AOT surfactants, the actual surface area per surfactant depends on the place of the benzyl sulfonate head group along the alkyl chain [33]. With increasing tail length, we also see that the optimum salinity shifts to lower values, again in good agreement with experiments [33].

After showing that the MoM is able to discriminate between different surfactant architectures and shows good agreement with experiments, we will finally focus on the role of the co-solvent.

13.5.3 EFFECT OF CO-SOLVENT

To further improve emulsification properties, usually a co-solvent such as butanol or pentanol is added to the surfactant mixture. In Table 13.4, we have collected MoM results obtained from DPD simulations of nonane–n10DTS–brine simulations, with and without 2% co-solvent isobutanol. To conform to the coarse-grained scheme we adopted earlier, isobutanol was modeled by two beads connected via a harmonic spring. Repulsive parameters were derived from a Flory–Huggins fitting procedure using force-field based molecular configurations and corresponding interaction energies. Figure 13.6 shows the simulated density profiles of the surfactant head, the surfactant tails, and the co-solvent. Although somewhat less pronounced, similar to the surfactant molecules, the co-solvent molecules aggregate at the oil–brine interface. Unlike the surfactant density profile, the co-solvent profile does not drop to zero in the oil and brine phase, indicative that the co-solvent will spread over a much large area of the interface in comparison with the surfactant. The presence of the co-solvent at the interface leads to a substantial increase in the average surface area per surfactant, which increases from 0.50 to 0.67 nm^2 [22]. The molar surface concentrations of the co-solvent and surfactant are approximately equal. For SDS–pentanol and SDS–hexanol mixtures, similar results were reported experimentally [34]. There is also a slight shift in optimum salinity to lower values, a fact that is

---

TABLE 13.4

Simulated Optimal Salt Strengths ($\log_{10} C(M)$) for the n10DTS Surfactant with and without Co-Solvent as Obtained from MoM Simulations

Surfactant	$\log_{10} C(M)$	A_{surf} (nm^2)	κ_s (10^{-21} Nm)
n10DTS	0.34	0.50	−9.9
n10DTS + isobutanol	0.30	0.67	−5.7

Source: M. Buijse et al., Accelerated surfactant selection for EOR using computational methods, SPE Enhanced Oil Recovery Conference, July 2–4, 2013, Kuala Lumpur, Malaysia, 165268-MS.

Note: A_{surf} is the surface area per surfactant at the interface at the optimum salinity and κ_s is the splay bending coefficient (or second moment of the stress) at optimum conditions ($\gamma = 0$, $\tau = 0$). The optimum salinity of n10DTS differs from value reported in Table 13.3 as, for the simulations here, a nonoptimized DPD parameter set was used.

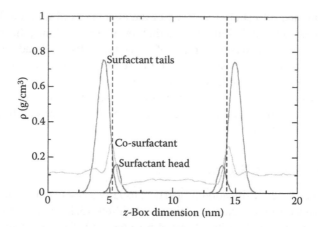

FIGURE 13.6 Density profiles of surfactant head, surfactant tails, and co-solvent obtained from a DPD simulation at optimum salinity using 2% co-solvent. Surfactant is n10DTS and co-solvent is iso-butanol. Density profiles of water, nonane, and salt ions are not visualized. The average positions of the two interfaces are indicated by the dashed lines. (Figure adapted from M. Buijse et al., Accelerated surfactant selection for EOR using computational methods, SPE Enhanced Oil Recovery Conference, July 2–4, 2013, Kuala Lumpur, Malaysia, 165268-MS.)

well known experimentally [22,34]. The second moment of the stress (κ_s), related to the bending rigidity of the surface film, indicates that the interface with co-solvent is more flexible than the interface without co-solvent. Clearly, population of isobutanol molecules at the interface and, conversely, depopulation of surfactant molecules lead to a less rigid surface film.

In recently published work, we discuss the possibility of a relation between the bending rigidity and the solubilization and viscosity of the microemulsion [22]. Preliminary results show that a decrease in the bending rigidity κ_s (i.e., the second moment of the stress) leads to a decrease in the viscosity of the microemulsion. We are currently investigating whether there is indeed a direct relation between the second moment of the stress and the microemulsion viscosity, in order to extend the screening capabilities of the MoM.

13.6 CONCLUDING REMARKS AND OUTLOOK

We have shown in this chapter that the combination of a coarse-grained particle-based simulation technique (i.e., DPD) and the MoM leads to a fast and accurate computational protocol for surfactant optimization in cEOR. The protocol has a great advantage that there is no need to perform simulations at ultralow IFT, which are inherently difficult to perform. Instead, we simply extrapolate to $\gamma = 0$. For neat oil–brine–surfactant interfaces, we could reproduce the optimum salinities found experimentally. Similarly, the change of the surfactant tail length and related shift of the optimum salinity could be quantitatively reproduced. The effect of the addition of a co-solvent (isobutanol) indicated a small shift to lower optimum salinity, while at the same time the interface becomes more flexible. It remains to be seen whether such a change in surface film rigidity can be linked to a change in microemulsion viscosity.

The simulation work presented in this chapter is a first step toward a robust and quantitative computational screening protocol for realistic applications in cEOR, which include "live" crude oils and surfactant mixtures. The second step would be to convert experimental data of crudes and surfactant mixtures into molecular models. Although, both on the theoretical and experimental side, a lot has been already accomplished (see, e.g., References 35–37), there is still a considerable task ahead to generate realistic molecular models from experimental data that is currently being generated by the oil industry and associated service companies. Once the molecular models of the crude and surfactant mixtures have been developed, one could do the MoM simulations to compute the properties of interest.

The actual formation of the microemulsion is, of course, only one part of cEOR. Currently, there is an ongoing effort by Shell and Schlumberger to develop simulation models to study flow in rock formations [38]. We should also not underestimate the problems associated with the breaking of the microemulsion at the production well [39].

ACKNOWLEDGMENTS

The authors would like to thank Shell Global Solutions International and the Culgi BV management for permission to publish this chapter.

REFERENCES

1. Y. Barakat, L.N. Fortney, C. LaLanne-Cassou, R.S. Schechter, W.H. Wade, U. Weerasooriya, and S. Yiv, The phase behavior of simple salt-tolerant sulfonates, *Soc. Petrol. Eng. J.* **1983**, *23*, 913–918.
2. R. Aveyard, B.P. Binks, and J. Mead, Interfacial tension minima in oil–water–surfactant systems. Effects of cosurfactant in systems containing sodium dodecyl sulphate, *J. Chem. Soc. Faraday Trans. I* **1987**, *83*, 2347–2357.
3. S. Karaborni, N.M. van Os, K. Esselink, and P.A.J. Hilbers, Molecular dynamics simulations of oil solubilization in surfactant solutions, *Langmuir* **1993**, *9*, 1175–1178.
4. G.J. Hirasaki, C.A. Miller, and M. Puerto, Recent advances in surfactant EOR, SPE Annual Technical Conference and Exhibition, September 21–24, 2008, Denver, SPE 115386.
5. J.-L. Salager, L. Manchego, L. Márquez, J. Bullón, and A. Forgiarini, Trends to attain a lower interfacial tension in a revisited pure alkyl polyethyleneglycol surfactant–alkane–water ternary system. Basic concepts and straightforward guidelines for improving performance in enhanced oil recovery formulations, *J. Surfact. Deterg.* **2014**, *17*, 199–213.
6. P.J. Hoogerbrugge, and J.M.V.A. Koelman, Simulating microscopic hydrodynamic phenomena with dissipative particle dynamics, *Europhys. Lett.* **1992**, *19*, 155–160.
7. P. Español, and P.B. Warren, Statistical mechanics of dissipative particle dynamics, *Europhys. Lett.* **1995**, *30*, 191–196.
8. R.D. Groot, and P.B. Warren, Dissipative particle dynamics: Bridging the gap between atomistic and mesoscopic simulation, *J. Chem. Phys.* **1997**, *107*, 4423–4435.
9. L. Rekvig, M. Kranenburg, J. Vreede, B. Hafskjold, and B. Smit, Investigation of surfactant efficiency using dissipative particle dynamics, *Langmuir* **2003**, *19*, 8195–8205.

10. V.V. Ginzburg, K. Chang, P.K. Jog, A.B. Argenton, and L. Rakesh, Modeling the interfacial tension in oil/water/nonionic surfactant mixtures using dissipative particle dynamics and self-consistent field theory, *J. Phys. Chem. B* **2011**, *115*, 4654–4661.

11. P. Carbone, H.A. Karimi-Varzaneh, and F. Müller-Plathe, Fine-graining without coarse-graining: An easy and fast way to equilibrate dense polymer melts, *Faraday Discuss.* **2010**, *144*, 25–42.

12. J.-W. Handgraaf, R. Serral Gracià, S.K. Nath, Z. Chen, S.-H. Chou, R.B. Ross, N.E. Schultz, and J.G.E.M. Fraaije, A multiscale modeling protocol to generate realistic polymer surfaces, *Macromolecules* **2011**, *44*, 1053–1061.

13. H.-J. Qian, C.C. Liew, and F. Müller-Plathe, Effective control of the transport coefficients of a coarse-grained liquid and polymer models using the dissipative particle dynamics and Lowe–Andersen equations of motion, *Phys. Chem. Chem. Phys.* **2009**, *11*, 1962–1969.

14. J.G.E.M. Fraaije, S.K. Nath, J. van Male, P. Becherer, J. Klein Wolterink, J.-W. Handgraaf, F. Case, C. Tanase, and R. Serral Gracià, *Culgi Manual*, version 8.0, Culgi B.V., the Netherlands, 2013.

15. P.G. de Gennes, and C. Taupin, Microemulsions and the flexibility of oil/water interfaces, *J. Phys. Chem.* **1982**, *86*, 2294–2304.

16. W. Helfrich, Elasticity and thermal undulations of fluid films of amphiphiles, in *Les Houches, Session XLVIII, 1988—Liquides aux Interfaces/Liquids at Interfaces*, Charvolin, J., Joanny, J.F., Zinn-Justin, J., Eds. Elsevier Science Publishers, Amsterdam, North-Holland, 1990.

17. I. Szleifer, D. Kramer, A. Benshaul, W.M. Gelbart, and S.A. Safran, Molecular theory of curvature elasticity in surfactant films, *J. Chem. Phys.* **1990**, *92*, 6800.

18. R. Strey, Microemulsion microstructure and interfacial curvature, *Colloid Polym. Sci.* **1994**, *272*, 1005–1019.

19. T. Sottmann, and R. Strey, Ultralow interfacial tension in water-n-alkane–surfactant systems, *J. Chem. Phys.* **1997**, *106*, 8606–8615.

20. S.A. Safran, Curvature elasticity of thin films, *Adv. Phys.* **1999**, *48*, 395–448.

21. M. Buijse, K. Tandon, S. Jain, J.-W. Handgraaf, and J.G.E.M. Fraaije, Surfactant optimization for EOR using advanced chemical computational methods, SPE Improved Oil Recovery Symposium, April 14–18, 2012, Tulsa, OK, 154212-MS.

22. M. Buijse, K. Tandon, S. Jain, A. Jain, J.-W. Handgraaf, and J.G.E.M. Fraaije, Accelerated surfactant selection for EOR using computational methods, SPE Enhanced Oil Recovery Conference, July 2–4, 2013, Kuala Lumpur, Malaysia, 165268-MS.

23. J.G.E.M. Fraaije, K. Tandon, S. Jain, J.-W. Handgraaf, and M. Buijse, Method of moments for computational microemulsion analysis and prediction in tertiary oil recovery, *Langmuir* **2013**, *29*, 2136–2151.

24. U. Essmann, L. Perera, M.L. Berkowitz, T. Darden, H. Lee, and L.G.A. Pedersen, Smooth particle mesh Ewald method, *J. Chem. Phys.* **1995**, *103*, 8577–8593.

25. G. Scocchi, P. Posocco, J.-W. Handgraaf, J.G.E.M. Fraaije, M. Fermeglia, and S. Pricl, A complete multiscale modelling approach for polymer–clay nanocomposites, *Chem., Eur. J.* **2009**, *15*, 7586–7592.

26. R. Toth, D.-J. Voorn, J.-W. Handgraaf, J.G.E.M. Fraaije, M. Fermeglia, S. Pricl, and P. Posocco, Multiscale computer simulation studies of water-based montmorillonite/poly(ethylene oxide) nanocomposites, *Macromolecules* **2009**, *42*, 8260–8720.

27. P. Becherer, Q. Dong, J.G.E.M. Fraaije, R. Serral Gracià, J. van Male, S.K. Nath, B. Vogelaar, Z. Wan, R. Wang, and Y. Yang, *The Chemistry Unified Language Interface (Culgi)*, versions 6.0 and 7.0, Culgi B.V., the Netherlands, 2004–2013.

28. R. Granet, R.D. Khadirian, and S. Piekarski, Interfacial tension and surfactant distribution in water–oil–NaCl systems containing double-tailed sulfonates, *Colloids Surf.* **1990**, *49*, 199–209.

29. G.J. Verhoeckx, P.L. Debruyn, and J.T.G. Overbeek, On understanding microemulsions. 1. Interfacial tensions and adsorptions of SDS and pentanol at the cyclohexane water interface, *J. Colloid Interface Sci.* **1987**, *119*, 409–421.

30. J.T.G. Overbeek, G.J. Verhoeckx, P.L. De Bruyn, and H.N.W. Lekkerkerker, On understanding microemulsions. II. Thermodynamics of droplet-type microemulsions, *J. Colloid Interface Sci.* **1987**, *119*, 422–441.

31. J.R. Barnes, H. Dirkzwager, J.P. Smit, J.R. Smit, Q.A. On, R.C. Navarette, B.H. Ellison, and M.A. Buijse, Application of internal olefin sulfonates and other surfactants to EOR. Part 1: Structure–performance relationships for selection at different reservoir conditions, SPE Improved Oil Recovery Symposium, April 24–28, 2010, Tulsa, OK, SPE-129766-MS.

32. R. Nagarajan, Molecular packing parameter and surfactant self-assembly: The neglected role of the surfactant tail, *Langmuir* **2002**, *18*, 31–38.

33. S.S. Jang, S.-T. Lin, P.K. Maiti, M. Blanco, W.A. Goddard III, P. Shuler, and Y. Tang, Molecular dynamics study of a surfactant-mediated decane–water interface: Effect of molecular architecture of alkyl benzene sulfonate, *J. Phys. Chem. B* **2004**, *108*, 12130–12140.

34. W.K. Kegel, G.A. van Aken, M.N. Bouts, H.N.W. Lekkerkerker, J.T.G. Overbeek, and P.L. de Bruyn, Adsorption of sodium dodecyl sulfate and cosurfactant at the planar cyclohexane–brine interface. Validity of the saturation adsorption approximation and effects of the cosurfactant chain length, *Langmuir* **1993**, *9*, 252–256.

35. E. Rogel, Theoretical estimation of the solubility parameter distributions of asphaltenes, resins, and oils from crude oils and related materials, *Energy Fuels* **1997**, *11*, 920–925.

36. E.S. Boek, M.R. Stukan, and P. Ligneul, Molecular dynamics simulation of surfactant flooding in asphaltenic oils for EOR applications, 16th European Symposium on Improved Oil Recovery, Cambridge, UK, April 12–14, 2011.

37. Y.E. Corilo, D.C. Podgorski, A.M. McKenna, K.L. Lemkau, C.M. Reddy, A.G. Marshall, and R.P. Rodgers, Oil spill source identification by principal component analysis of electrospray ionization Fourier transform ion cyclotron resonance mass spectra, *Anal. Chem.* **2013**, *85*, 9064–9069.

38. R. Armstrong, N. Evseev, O.M. Gurpinar, L.A. Hathon, D.V. Klemin, O. Dinariev, S. Berg, S. Safonov, M. Myers, H. de Jong, C. van Kruijsdijk, and D.A. Koroteev, Application of digital rock technology for chemical EOR screening, SPE Enhanced Oil Recovery Conference, July 2–4, 2013, Kuala Lumpur, Malaysia, SPE-165258-MS.

39. T.N. de Castro Dantas, A.A. Dantas Neto, and E. Ferreira Moura, Microemulsion systems applied to breakdown petroleum emulsions, *J. Petrol. Sci. Eng.* **2001**, *32*, 145–149.

14 Understanding the Molecular Information Contained in the Infrared Spectra of Colombian Vacuum Residua by Principal Component Analysis

Jorge A. Orrego-Ruiz, Daniel Molina,
Enrique Mejía-Ospino, and Alexander Guzmán

CONTENTS

CONTEXT

- Use of chemometric techniques to characterize petroleum fractions
- Prediction of physical properties on the basis of spectroscopic data

ABSTRACT

Infrared (IR) spectroscopy together with chemometric approaches was used to obtain detailed molecular information on eight vacuum residua and their molecular distillation fractions. Both hierarchical cluster analysis and principal component analysis done from two IR regions, 2800–3000 cm^{-1} and 700–900 cm^{-1}, led to the proposal of four new structural parameters. This information can be examined to propose new mixing schemes of heavy residues in the refinery. To validate the structural information given by these parameters, micro carbon residue, density, and hydrogen/carbon ratio from 21 samples were obtained, as well as the ^{1}H and ^{13}C-NMR spectra.

14.1 INTRODUCTION

In the last decade, Colombia has increased its oil production to up to 1 million barrels a day [1]. Because Colombian crude oils exhibit a significant compositional variety [2], it requires a judicious blending according to physicochemical properties such as density, sulfur, and metal contents; micro carbon residue (MCR); viscosity; yield curve measured by simulated distillation; and saturates, aromatics, resins, and asphaltene contents (SARA analysis). Importantly, the quality of the final products depends on a successful blending process. However, conventional characterization of heavy crudes (whose reserves are steadily increasing [3]) that are being incorporated in the crude oil diets of the two major refineries in the country could be insufficient to produce optimum yields in some processes. Several studies have been reported regarding molecular characterization of crude oils and vacuum residues using gas chromatography [4,5], mass spectrometry [6,7], fluorescence [8], infrared (IR) [9], and nuclear magnetic resonance (NMR) spectroscopies [10–12]. Nevertheless, heavy crudes—and more specifically their residues—are sufficiently complex that several of these techniques must be engaged for a full characterization. As it is well known, vacuum residua (VR), the nondistillable fraction of vacuum distillation, concentrate resins and asphaltenes. VR are complex mixtures of thousands of molecules [13] and require conversion processes such as thermal cracking or delayed coking to become valuable products [14–16]. In general, many of the molecules in VR show high aromaticity or high molecular weight as a natural consequence of being the distillation bottom under reduced pressure. This fact limits the compositional analysis using techniques successfully used with light and middle distillate fractions, such as gas chromatography. On the other hand, although techniques like NMR or Fourier transform ion cyclotron mass spectrometry are available for the molecular characterization of VR, the interpretation of the results is often very difficult. One way to reduce complexity is the fractionation of VR by supercritical fluid extraction [17], solvent extraction [18–20], or short-path distillation (molecular distillation). The latter is a fractionation process that uses high vacuum (around 1×10^{-2} Torr), temperatures <350°C, and a special design for the evaporator to reduce material exposure time avoiding thermal decomposition of the feed [21]. By using these conditions, it is possible to achieve distillation temperatures up to 689°C AET (atmospheric temperature equivalent). This process has found important applications in the purification of heat-sensitive mixtures [22] and in the extension of the true boiling point curve for crude oils from 510°C up to 690°C [23–27].

The present work shows the characterization of eight VR along with their molecular distillation fractions using IR spectroscopy and chemometrics. While IR spectroscopy generally is limited to the determination of functional groups, analysis of IR spectra using chemometrics allows obtaining valuable structural information. The final result of this analysis is the method to obtain four structural parameters from which it is possible to calculate the density and MCR of heavy fractions. Moreover, structural information from the four structural parameters can be used as an alternative to monitor deep conversion processes of heavy fractions, or to be included with the conventional analysis to propose new mixing schemes of heavy residues in the refinery.

14.2 EXPERIMENTAL

14.2.1 MATERIALS AND METHODS

14.2.1.1 Samples

Eight VR from typical Colombian crude oils were used in this study. They were obtained following the standard distillation methods ASTM 2892 and ASTM 5236-06. All VR were fractionated using short-path distillation with a KD-6-15 distiller from Chemtech Services. Table 14.1 reports the yields of the cuts IBP-603°C, IBP-645°C, and IBP-687°C, which were named 2, 4, and 6, respectively. For example, the cut IBP-603°C from the vacuum residue R was labeled C4R, while its residue (603°C+) was labeled R4R.

14.2.1.2 Characterization of Samples

VR were characterized by API gravity (ASTM D70), sulfur content (ASTM D1552), MCR (ASTM D4530), and SARA analysis (ASTM D2549/ASTM D2007 and IP-143). These values are reported in Table 14.1. Density and MCR varied gradually from heavy (sample C) to light (sample T) samples.

Additionally, VR R, G, and T along with their fractions were characterized by C and H elemental content (ASTM D 5291), API gravity, sulfur content, and ^{13}C- and ^{1}H-NMR spectroscopy (Table 14.2). The choice of these samples was done according to their provenance—R comes from a heavy crude oil, while G and T come from a medium and a light crude oil, respectively—to obtain a good variability of properties. NMR assignments and band integrations were done following the procedure of Poveda [28] and Molina [29]. Table 14.2 is given in Section 14.3.5 after explaining the meaning and the origin of the structural parameters proposed, for a better understanding of the results.

TABLE 14.1

Physicochemical Characterization and Molecular Distillation Yield of Vacuum Residues

Property	Units	C	R	V	CL	Te	SF	G	T
Density	°API	0.8	2.0	3.4	4.7	4.7	7.1	9.9	11.6
Sulfur	wt.%	3.60	2.05	2.12	1.63	2.78	1.43	1.85	2.20
MCR[a]	wt.%	35.1	35.1	28.6	27.3	25.1	21.5	18.3	18.3
Saturates	wt.%	4.1	13.0	5.4	11.5	6.4	16.5	16.7	21.7
Aromatics	wt.%	37.1	39.5	40.7	36.0	45.7	41.7	50.9	50.8
Resins	wt.%	27.5	18.2	35.8	33.8	30.4	30.2	26.0	21.4
Asphaltenes	wt.%	31.3	29.3	18.1	18.6	17.4	11.6	6.4	6.1
Sat + arom	wt.%	41.2	52.5	46.1	47.5	52.1	58.2	67.6	72.5
IBP-603°C	wt.%	2.2	6.1	12.8	29.3	18.1	37.2	21.4	21.2
IBP-645°C	wt.%	7.8	17.7	38.6	42.5	25.9	47.7	33.6	39.8
IBP-687°C	wt.%	21.8	34.4	46.1	54.8	33.8	56.2	44.5	60.0

Note: The names used (C, R, V, etc.) to describe vacuum residues are fictitious and are not related to the origin of the samples.

[a] Corresponds to micro carbon residue.

14.2.2 ACQUISITION OF IR SPECTRA

Fourier transform IR spectra in attenuated total reflection (ATR) mode were recorded on an IR-Prestige 21 (Shimadzu) spectrometer with a spectral resolution of 8 cm^{-1} over the range of 4000–650 cm^{-1} and 32 scans. The spectrometer was equipped with a deuterated triglycine sulfate detector and a Pike Miracle ATR diamond cell with a single reflection (45°). An adjustable pressure system was used for solid samples; meanwhile, liquid samples were spread on the crystal. After measurement of the spectra, the remaining sample on the crystal was removed with a soft solvent wet paper. Background was collected for every sample before the acquisition with the same clean diamond cell. The final IR spectrum of each sample was the average result of three acquisitions.

14.2.3 ACQUISITION OF ^{1}H- AND ^{13}C-NMR SPECTRA

^{1}H- and ^{13}C-NMR spectra were obtained with a spectrometer (Bruker Avance III) at 400.16 and 100.62 MHz, respectively. For ^{1}H-NMR, the acquisitions were carried out using sample solutions to 4 wt.% in CDCl$_3$ (99.8% D). For each spectrum, 30° pulses (Bruker zg30 pulse sequence) and a delay time of 10 s (sweep width 6000 Hz) were used; 16 scans were averaged. For ^{13}C-NMR, spectra were recorded in a 5-mm probe and at a temperature of 298.15 K using sample solutions to 20 wt.% in CDCl$_3$ (99.8% D), and 0.05 M Cr(acac)$_3$ as a paramagnetic relaxation reagent. ^{13}C-NMR spectra were acquired using 30° pulses (Bruker zgig30 pulse sequence) and a delay time of 20 s (sweep width 24040 Hz); 1024 scans were averaged for each spectrum. Spectra were recorded in a 10-mm probe with a spinning rate of 30 Hz and a temperature of 298.15 K.

TABLE 14.2
Physicochemical Characterization of R, G, and T Vacuum Residues and Their Fractions

	Sulfur[a]	Carbon[a]	Hydrogen[a]	H/C	MCR[a,b]	Density[c]	AC[d] (IR)	LACAR[e] (IR)	AI[f] (IR)	ACL[g] (IR)	f_a^h (^{13}C RMN)	CH$_2$/CH$_3$ (^{13}C-RMN)	Area 1–1.5 ppm (^{1}H RMN)
C2T	1.47	85.93	12.43	1.72	5.7	19.3	0.68	0.45	0.41	2.26	0.14	0.7	57.75
C4T	1.54	85.7	12.16	1.69	5.5	19.2	0.72	0.48	0.40	2.33	0.17	0.85	58.06
C6T	1.53	86.1	12.52	1.73	8.3	17.7	0.73	0.49	0.31	2.46	0.16	0.73	56.33
T	2.12	85.35	11.61	1.62	18.3	11.6	0.97	0.30	0.33	2.25	0.27	0.71	56.54
R2T	2.23	85.88	11.21	1.56	19.4	11.9	1.06	0.29	0.34	2.37	0.23	0.68	49.23
R4T	2.47	86.1	11.25	1.56	24.7	9.2	1.16	0.26	0.33	2.39	0.25	0.71	54.04
R6T	2.73	85.47	10.15	1.42	29.5	5.7	1.26	0.21	0.26	2.31	0.31	0.68	49.87
C2G	1.41	85.71	11.76	1.63	4.6	15.5	0.77	0.25	0.36	1.77	0.18	0.41	49.53
C4G	1.39	85.83	11.89	1.65	5.3	15.7	0.80	0.25	0.38	1.82	0.20	0.48	51.22
C6G	1.41	86.03	12.16	1.68	7.2	15.1	0.80	0.26	0.31	1.83	0.15	0.55	51.26
G	1.77	85.98	10.88	1.51	18.3	9.9	1.06	0.22	0.27	1.94	0.24	0.55	48.97
R2G	1.91	86.22	10.73	1.48	23.6	7.6	1.16	0.21	0.33	1.94	0.25	0.63	46.81
R4G	2.00	85.5	10.72	1.49	27.1	6.0	1.13	0.21	0.29	1.97	0.25	0.59	48.39
R6G	2.07	85.7	10.18	1.42	29.2	4.6	1.24	0.18	0.30	1.99	0.27	0.61	46.89
C2R	1.49	86.96	11.04	1.51	10.4	10.2	0.93	0.15	0.31	1.57	0.22	0.33	45.07

(Continued)

TABLE 14.2 (Continued)

Physicochemical Characterization of R, G, and T Vacuum Residues and Their Fractions

	Sulfur[a]	Carbon[a]	Hydrogen[a]	H/C	MCR[a,b]	Density[c]	AC[d] (IR)	LACAR[e] (IR)	Al[f] (IR)	ACL[g] (IR)	$f_a^{[h]}$ (^{13}C RMN)	CH$_2$/CH$_3$ (^{13}C-RMN)	Area 1–1.5 ppm (^{1}H RMN)
C4R	1.52	86.76	11.61	1.59	8.7	10.7	0.90	0.15	0.34	1.59	0.20	0.38	43.7
C6R	1.45	86.52	11.81	1.63	9.0	10.5	0.92	0.15	0.33	1.61	0.22	0.39	45.77
R	2.11	87.34	9.52	1.3	35.1	2.0	1.31	0.12	0.26	1.69	0.38	0.4	41.59
R2R	2.15	86.52	9.02	1.24	38.0	0.4	1.36	0.12	0.25	1.72	0.36	0.41	41.45
R4R	2.23	86.98	9.04	1.24	41.0	−0.5	1.36	0.11	0.24	1.71	0.38	0.38	41.85
R6R	2.29	86.96	8.6	1.18	44.5	−2.1	1.36	0.11	0.23	1.67	0.39	0.38	41.22

[a] Values measured in wt.%.

[b] Micro carbon residue.

[c] Values measured in API.

[d] Aromatic condensation by Equation 14.2.

[e] Long aliphatic chains over aromatic rings by Equation 14.3.

[f] Aliphaticity by Equation 14.4.

[g] Aliphatic chain length by Equation 14.5.

[h] Aromaticity factor according to Poveda [34].

14.2.4 PRINCIPAL COMPONENT ANALYSIS AND HIERARCHICAL CLUSTER ANALYSIS

Before chemometric analyses, all spectra were transformed through baseline and area normalization pretreatments; finally, they were mean centered. Principal component analysis (PCA) was carried out with the Unscrambler software version 10.2 by using the NIPALS (nonlinear iterative partial least squares) algorithm. Hierarchical cluster analysis (HCA) was performed according to the average linkage clustering using absolute correlation. Importantly, PCA helps determine in what aspect samples differ from each other, which spectral intensities contribute most to this difference, and whether those intensities are correlated or not. It enables the detection of sample patterns, and quantifies the amount of useful information contained in the data. PCA can be defined as an orthogonal linear transformation that transforms the data to a new coordinate system such that the greatest variance by any projection of the data comes to lie on the first coordinate (called the first principal component), the second greatest variance on the second coordinate, and so on. A deeper discussion about the mathematics behind these concepts is out of the scope of this chapter.

On the other hand, HCA can be defined as a data analysis technique applied to a set of heterogeneous items that identifies homogeneous subgroups as defined by a measure of similarity [22]. Since each sample is described by hundreds of IR intensities, similarity measures can be based on the Pearson correlation coefficient (r). The Pearson correlation coefficient can be calculated according to Equation 14.1, where x_i and y_i correspond to the ith IR intensity of the samples x and y, n is the total number of IR intensities, $\bar{x}$ and $\bar{y}$ correspond to the average of IR intensities, and s_x and s_y are the standard deviations. This coefficient takes values from -1 (large negative correlation) to $+1$ (large positive correlation). Unlike PCA, this analysis detects global similarity between samples because it takes into account the average of intensities and does not transform the original data as PCA does.

$$r = \frac{\sum_{i=1}^{n}(x_i - \bar{x})(y_i - \bar{y})}{\sqrt{\sum_{i=1}^{n}(x_i - \bar{x})^2 \sum_{i=1}^{n}(y_i - \bar{y})^2}} = \frac{\text{Cov}(x, y)}{(n-1)s_x s_y} \qquad (14.1)$$

14.3 RESULTS AND DISCUSSION

14.3.1 CHARACTERIZATION OF SAMPLES

According to the results in Table 14.1, no tendency between physicochemical characterization from VR and their molecular distillation yields was found. For example, crude CL (4.7°API) had higher yields for IBP-603°C and IBP-645°C than crude T (11.6°API), even though CL—with respect to T—has a lower sum of saturates plus aromatics (S + Ar), which is directly related to the distillable fraction. It means that the physicochemical properties listed in Table 14.1 do not satisfactorily explain the distillation yields.

14.3.2 IR Analysis

Concerning petroleum-derived hydrocarbons, there are two IR regions where functional groups appear, i.e., 650–1800 cm^{-1} (bending vibration modes) and 2650–3100 cm^{-1} (stretching vibration modes). The IR spectra for all the samples had a similar appearance. Assignments of the main bands are presented in Figure 14.1 [30]. For the same functionality, it is possible to find several assignments spread out in the mid-IR region. For instance, methyl (CH$_3$) groups can be detected at 2954 and 2894 cm^{-1} because of their asymmetric and symmetric stretching and also at 1463 and 1376 cm^{-1} because of their asymmetric and symmetric bending. This means four bands can describe the same CH$_3$ moiety. On the other hand, the spectral intensities depend on the concentration of the functional groups, the molar absorptivity, and the optical path length of the IR radiation. The main difficulty in quantifying the species by IR spectroscopy lies in not knowing the absorptivity values, because for two species with the same concentration, the intensity for the lower absorptivity functionality will be lower than that found for the higher absorptivity analog. However, assuming that the absorptivity values do not change significantly between samples with the same nature, the variations in IR intensities between VR and their fractions can be used to explain the behavior during molecular distillation. For that, we use chemometrics.

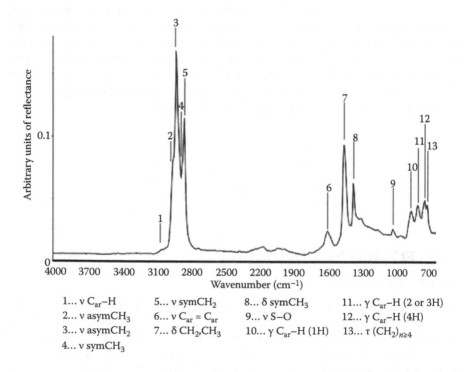

1... ν C$_{ar}$–H 5... ν symCH$_2$ 8... δ symCH$_3$ 11... γ C$_{ar}$–H (2 or 3H)
2... ν asymCH$_3$ 6... ν C$_{ar}$ = C$_{ar}$ 9... ν S–O 12... γ C$_{ar}$–H (4H)
3... ν asymCH$_2$ 7... δ CH$_2$,CH$_3$ 10... γ C$_{ar}$–H (1H) 13... τ (CH$_2$)$_{n≥4}$
4... ν symCH$_3$

FIGURE 14.1 MIR–ATR spectrum of vacuum residues.

14.3.3 HIERARCHICAL CLUSTER ANALYSIS

HCA was performed according to average linkage clustering [31] through absolute correlation using both aromatic and aliphatic IR spectral regions: 700–900 cm^{-1} and 2835–3000 cm^{-1}. The main outcome from HCA is shown through the dendogram in Figure 14.2, where five clusters were found. These clusters can be grouped as follows:

1. *Extra-heavy residues*: Corresponding to three residues from C.
2. *Heavy residues*: This cluster includes the residues from Te, V, R, CL, and the vacuum residues V, R, and C.
3. *Light condensates*: This cluster includes the condensates from SF, CL, and the vacuum residue T together with its residues.
4. *Medium residues*: This cluster includes the residues from SF, G, and the vacuum residues Te, G, SF, and CL.
5. *Heavy condensates*: It includes the condensates C, G, V, R, and Te.

Figure 14.2 shows that there are two main branches. The first one is formed by clusters 1 and 2, and the second one by clusters 3, 4, and 5. It means that residues from vacuum residues C, R, Te, V, and CL have a high similarity. Likewise, the SF and G residues have a high similarity. VR were circled in red to distinguish them from among all samples and were grouped together with their residues. They were clustered according to API gravity: C, R, and V within the heavy residues in cluster 2; CL, Te, SF, and G within the medium residues in cluster 4; and T within the light condensates in cluster 3. Cluster 3 grouped condensates of highest yields (SF, T, and CL). The details of such functional groups can be seen through PCA.

14.3.4 PRINCIPAL COMPONENT ANALYSIS

Two PCAs were performed using aromatic and aliphatic spectral regions: 700–900 cm^{-1} and 2835–3000 cm^{-1}, respectively. In both cases, the first three principal components (PCs) explained >96% of the total variance. A total of 56 samples including all the VR and fractions were employed in the analyses. To determine the structure in the IR intensities (*X*-variables) and to demonstrate their significance, correlation loading plots were examined. Figure 14.3 presents the loading plots for aromatic and aliphatic regions. Each plot in Figure 14.3 contains two ellipses that indicate how much variance is taken into account. The outer ellipse is the unit circle and indicates 100% of the explained variance. The inner ellipse indicates 50% of the explained variance. Most IR intensities were inside the outer ellipse for both spectral regions. This indicates that they contain enough structured variation to be discriminators of the samples. In Figure 14.3a, it is clear that there are two regions that vary in a juxtaposed way, and there is no clear separation between them. For that reason, the intensities 736, 732, and 729 cm^{-1} are inside of the inner ellipse.

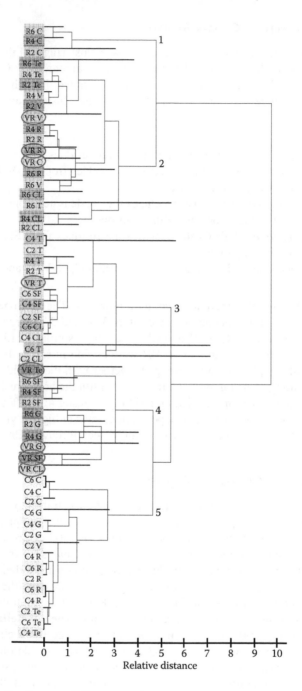

FIGURE 14.2   Dendogram of IR spectra of samples.

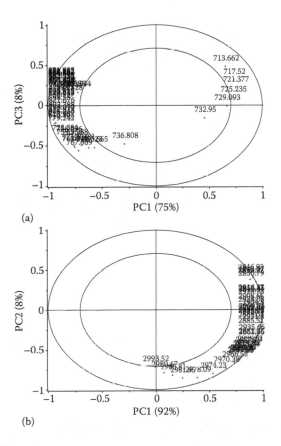

FIGURE 14.3 Loading map for (a) aromatic region and (b) aliphatic region.

14.3.4.1 PCA for the Aromatic Region, 710–900 cm⁻¹

Three PCs explained 97% of the data variance (PC1, 75%; PC2, 14%; and PC3, 8%), and none of the samples were considered outliers. Since the purpose of our work was to get a physical sense of the results, PC2 was not taken into account even though it explained 14% of the variance. The loading plot (not included) showed that PC2 is related to scattering of radiation, showing an increasing of the baseline and not with some specific functional groups, so much so that the score plot of PC1 vs. PC2 did not cluster samples reasonably. Nevertheless, some similar patterns in the dendogram from HCA (Figure 14.2) were detected when PC1 and PC3 were plotted. Figure 14.4 recaps the results of the PCA in the aromatic region. In the score map (Figure 14.4a), the condensates are along quadrants III, IV, and I in decreasing order of density: the heavy, medium, and light ones, and the residues are concentrated in quadrants I and II. This grouping can be explained from the loading plots of PC1 and PC3 (Figure 14.4b and c). Loadings describe the data structure in terms of variable contributions and correlations. Each X-variable (IR intensity) has a loading on each PC, which reflects how much the individual X-variable contributes to each PC, and

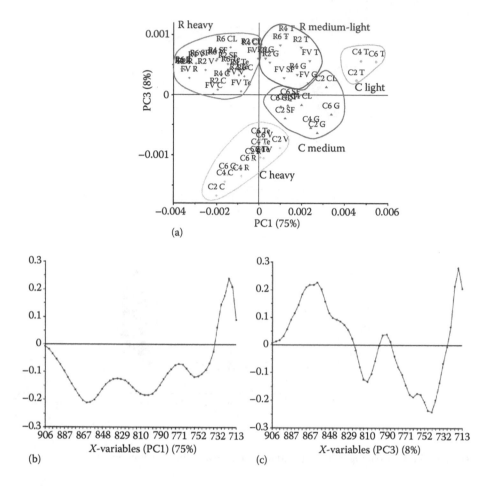

FIGURE 14.4 Results from the PCA in the aromatic region: (a) score map; (b) loadings for PC1; and (c) loadings for PC3.

how well the PC takes into account the variation in each X-variable. The positive loadings explain the position of the samples in positive scores and vice versa.

Figure 14.4b shows that PC1 is related to the relationship between long aliphatic chains and aromatic systems because the band attributed to rocking aliphatic in chains with more than three adjacent $-CH_2-$ (τ CH_2; assignment 13 in Figure 14.1) increases, while the aromatic triplet (γ $C_{ar}-H$ (1H), γ $C_{ar}-H$ (2H or 3H), and γ $C_{ar}-H$ (4H); assignments 10, 11, and 12 in Figure 14.1) have negative values.

On the other hand, from Figure 14.4d, it is inferred that PC3 is related to the aromatic condensation (AC) of the compounds because the band at 870 cm^{-1} (attributed to polysubstituted aromatic rings) is increasing at the expense of the band at 748 cm^{-1} (attributed to monosubstituted aromatic rings for PC3). Nevertheless, in PC3, not only the polysubstituted aromatic band has positive loadings but also the band attributed to rocking aliphatic in chains. That is the reason why both residues

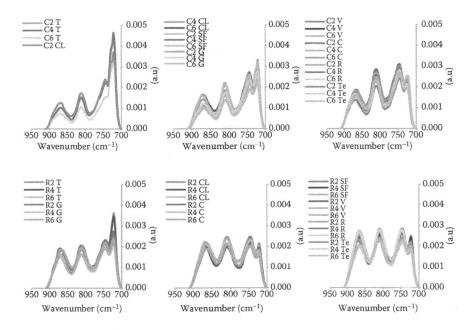

FIGURE 14.5 Spectra of samples within the IR aromatic region (710–900 cm^{-1}).

and light condensates have positive scores for PC3. These results confirm the effectiveness of IR spectroscopy in discriminating samples from different origins through PCA [32,33].

After the previous considerations, the spectra of the pooled samples on each cluster were examined (Figure 14.5). According to Figure 14.5, condensate bands showed similarities detected by HCA and PCA. The condensates T and C2CL are different with respect to the remaining group due to the band at 720 cm^{-1}. C4CL, C6CL, and condensates SF and G have different IR profiles with respect to condensates C, R, V, and Te, whose band at 810 cm^{-1} is more intensive. Residues T and G are clearly differentiated according to the band at 720 cm^{-1} from the other group of residues, as Figure 14.4a shows. The relative intensities of the four bands observed in the aromatic region, together with the loading plots (Figure 14.4c and d), suggest two structural parameters, which are discussed in the following.

14.3.4.2 PCA for the Aliphatic Region, 2835–3000 cm^{-1}

Two PCs were used to explain 99% of the data variance (PC1, 91.0% and PC2, 8.0%), and none of the samples were considered outliers. Patterns were detected when PC1 and PC2 were plotted. Figure 14.6 summarizes the mean results for the PCA in the aliphatic region. In the score map (Figure 14.6a), samples were discriminated to light and heavy condensates, and residues. Within light condensates, R2T and R4T were included. Unlike PCA in the aromatic region, condensates were grouped only into two sets. Most residues had negative values in PC1, while all condensates had positive values. PC2 discriminated condensates into light and heavy ones. According to the loading plot (Figure 14.6b and c), PC1 is related to the entire aliphatic region

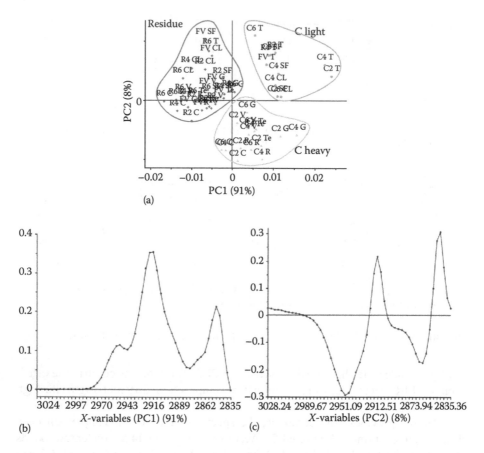

FIGURE 14.6   Results from the PCA in the aliphatic region: (a) score map; (b) loadings for PC1; and (c) loadings for PC3.

2800–3000 cm⁻¹, i.e., with aliphaticity. Samples with high aliphatic content, such as condensates, have positive values for PC1, while samples with low content of aliphatic species have negative values.

On the other hand, PC2 suggests important differences between samples regarding the aliphatic chain length (ACL). Asymmetric and symmetric ν–CH$_2$– stretching (assignments 3 and 5 in Figure 14.1) had positive values, while asymmetric and symmetric ν–CH$_3$– stretching (assignments 2 and 4 in Figure 14.1) had negative values for PC2. This means that distillation of T, SF, and CL provides condensates with a similar ACL with respect to their respective residues, whereas distillation of G, Te, V, R, and C provides condensates with shorter aliphatic chains (which may imply naphthenic compounds) with respect to their respective residues. Finally, samples were better clustered in the region 700–900 cm⁻¹ (five clusters) than in the region 2800–3000 cm⁻¹ (three clusters).

14.3.5 STRUCTURAL PARAMETERS

According to the previous analysis, correlations for four structural parameters emerge (Equations 14.2 through 14.5). These parameters were calculated for the VR and their fractions. First, a positive correlation between ACL and the yields (Figure 14.7) was detected. VR with higher ACL had higher yields and vice versa. How is it possible to explain why vacuum residue CL with API 4.7 had yields as high as T with API 11.0, or that Te with API 4.7 had a lower yield than V with 3.4? This can be explained by proposing that SF and CL have high concentrations of compounds distillable at 603°C, 645°C, and 687°C with long aliphatic chains, and a highly condensed nondistillable aromatic fraction at these temperatures. The AC values from Figure 14.8a show that SF and CL residues had unexpected AC values, ranking among the heaviest residues (R, V, and C). Likewise, the ACL values for their condensates showed that they have a high concentration of long aliphatic chains (Figure 14.8b), ranking among the lightest condensates (T).

$$\text{Aromatic condensation (AC)} = \frac{A_{825-906 \text{ cm}^{-1}}}{A_{736-783 \text{ cm}^{-1}}} = \frac{\gamma \; C_{ar} - H \, (1H)}{\gamma \; C_{ar} - H \, (4H)} \qquad (14.2)$$

$$\text{Long aliphatic chains over aromatics rings (LACAR)} = \frac{A_{713-732 \text{ cm}^{-1}}}{A_{736-902 \text{ cm}^{-1}}}$$

$$= \frac{\tau(CH_2)_{n \geq 4}}{\gamma \; C_{ar} - H \, (1-4H)} \qquad (14.3)$$

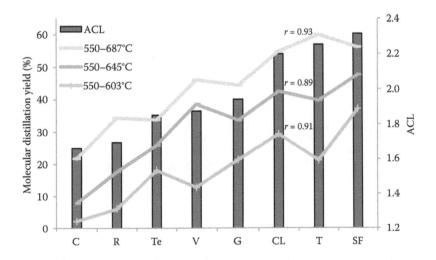

FIGURE 14.7 Correlations between ACL and distillation yields.

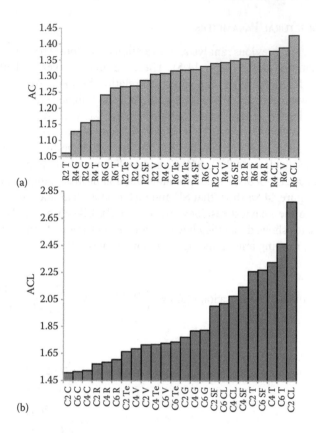

(a)

(b)

FIGURE 14.8 (a) AC values for residues and (b) ACL values for condensates.

$$\text{Aliphaticity (A1)} = A_{2835-2993\ cm^{-1}} = (\upsilon CH_2 + \upsilon CH_3) \tag{14.4}$$

$$\text{Aliphatic chain length (ACL)} = \frac{A_{2843-2854\ cm^{-1}} + A_{2908-2924\ cm^{-1}}}{A_{2862-2874\ cm^{-1}} + A_{2943-2959\ cm^{-1}}}$$

$$= \frac{\upsilon\ asymCH_2 + \upsilon\ symCH_2}{\upsilon\ asymCH_3 + \upsilon\ symCH_3} \tag{14.5}$$

However, the validity of all parameters was assessed using a detailed characterization of vacuum residues R, G, and T, and their fractions (Table 14.2). Variation between the H/C ratio and aliphaticity (A1) and AC can be seen in the Figure 14.9. As expected, correlations were positive for A1 and negative for AC. However, samples C6G and C6T did not show that trend.

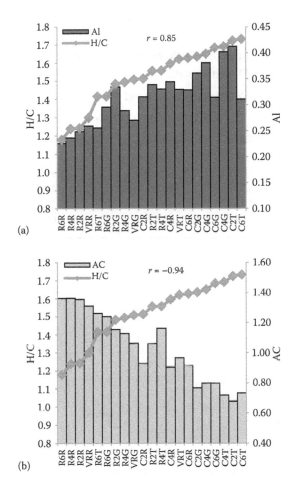

FIGURE 14.9 Correlation between elemental analysis (H/C) and (a) Al and (b) AC.

14.3.5.1 Correlation between Structural Parameters and Density

Parameters Al, AC, and ACL were correlated with the density of vacuum residues R, T, and G and their fractions through a multiple linear regression (MLR). The analysis of variances (ANOVA) is shown in the Table 14.3. Both F-ratio and p-value showed that the variance of the density can be explained through the variance of structural parameters. Using B-coefficients, a regression model can be written as shown in Equation 14.6. This equation confirms that the parameters of Al and ACL have positive correlations with API, whereas AC has a negative correlation. Figure 14.10 shows the residual and scattering plots. The R^2 (0.985) from Table 14.3 corresponds to a regression coefficient of scattering plot from Figure 14.10a. This means that >98% of variance in API can be explained through the model. On the other hand, Figure 14.10b shows a normal distribution of errors along the API values. These results demonstrate the significance of the MLR model.

$$API = 16.8 + 17.2x(A1) - 21.2x(AC) + 4.66x(ACL) \qquad (14.6)$$

TABLE 14.3

Analysis of Variances (ANOVA) between the Structural Parameters Al, AC, ACL and Density (°API)

Multiple correlation: 0.992 (cal); 0.989 (val)

R^2: 0.985 (cal); 0.977 (val)

	F-Ratio	p-Value	B-Coefficients	t-Values
Model	365	0.000		
Offset	19.3	0.000	16.8	4.4
Al	6.54	0.020	17.2	2.6
AC	170	0.000	−21.2	−13.1
ACL	50.9	0.000	4.66	7.1

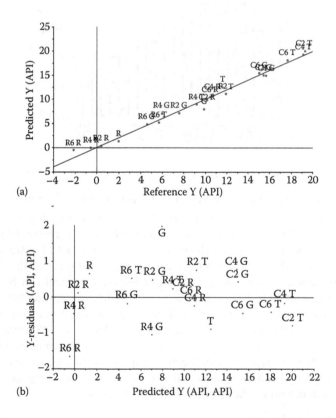

FIGURE 14.10 Plots of (a) scattering and (b) residuals of model to predict API.

14.3.5.2 Correlation between MCR and Structural Parameters

Similarly, parameters LACAR (long aliphatic chains associated with aromatic rings), AC, and ACL were used to predict MCR. The model was built with 20 samples. By excluding C4T, the model improved notably. Table 14.4 shows the ANOVA for the MLR. This model was significant in terms of high F-ratio and low p-value. Nevertheless, the model had a lower R^2 and a higher distribution of errors along the MCR values (Figure 14.11) in comparison with the API model. Figure 14.11b shows a high dispersion in the errors for the residues and VR (T and G) with respect to condensates. This can be explained by considering that the residues and VR had important asphaltene contents, which are important coke precursors. Therefore, a small increase in the asphaltene content means a high increase in MCR, which was not necessarily observed with density. It also means that the correlation between molecular structure and MCR depends strongly on the presence of asphaltenes. The MLR model can be written as shown in Equation 14.7. From Equation 14.7, it is possible to determine that the parameters related to aromaticity, LACAR, and AC have positive correlations with MCR, while ACL has a negative correlation. As it is well known, compounds with high aromatic content and short aliphatic moieties are more prone to coke formation.

$$\text{MCR (\%)} = -46.4 + 86.7x(\text{LACAR}) + 78.4x(\text{AC}) - 18.0x(\text{ACL}) \qquad (14.7)$$

Equations 14.6 and 14.7 were validated using 16 heavy residues. API and MCR were measured following the standard methods ASTM D-70 and ASTM-4530, respectively. For these samples, the four structural parameters AC, LACAR, Al, and ACL (Equations 14.2 through 14.5) were calculated using another IR instrument and following the same procedure of this work. Figure 14.12 shows the scattering plots to the API and MCR. High correlations were obtained for both properties ($R^2 = 0.968$ to API and 0.988 to MCR, respectively) in the 1–16° range for API and 0–30% range for MCR. Although the slopes and the offset must be corrected to use Equations 14.6 and 14.7 for other equipment, the high R^2 in both cases demonstrates once more the validity of the results, which was the main purpose in this work. It means that

TABLE 14.4
Analysis of Variances (ANOVA) between the Structural Parameters Al, AC, ACL, and MCR

Multiple correlation: 0.989 (cal); 0.984 (val)

R^2: 0.978 (cal); 0.969 (val)

	F-Ratio	p-Value	B-Coefficients	t-Values
Model	240	0.000		
Offset	120	0.000	−46.4	−11.0
LACAR	30.1	0.000	86.7	5.5
AC	278	0.000	78.4	16.7
ACL	19.6	0.000	−18.0	−4.4

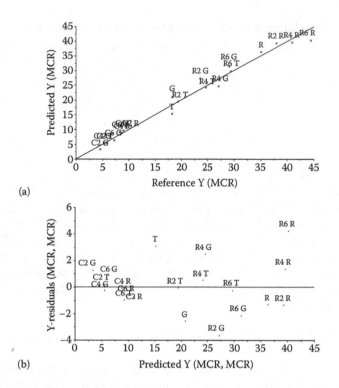

(a)

(b)

FIGURE 14.11   Plots of (a) scattering and (b) residuals of model to predict MLR.

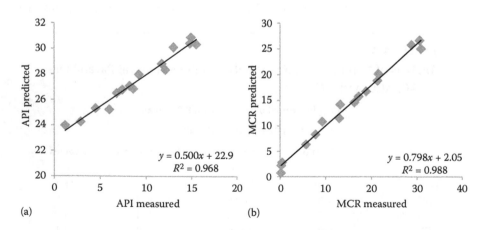

(a)

(b)

FIGURE 14.12   Plots of correlation (a) API and (b) MCR using a second IR equipment.

variances in density and MCR could be explained almost entirely by the four structural parameters proposed in Equations 14.2 through 14.5.

14.3.6 NMR ANALYSIS

Proton and carbon NMR of samples showed similar profiles that are only differentiated by their relative intensities. Both ^{1}H- and ^{13}C-NMR spectra from sample C2G are shown in Figures 14.13 and 14.14 along with their assignments [33]. ^{1}H-NMR spectra were integrated into six regions, as shown in Figure 14.13. Each region can be used to determine the concentration of functional groups where hydrogen is involved. In the case of ^{13}C-NMR, spectra were integrated into seven different regions (Figure 14.14). Region number 3 was additionally integrated into the 30.30–29.75 ppm range (region 3*). This region is characteristic of methylene moieties present in long aliphatic chains [34]. ^{1}H- and ^{13}C-NMR integrals were correlated with the structural parameters. The best results are presented in the following. Aromaticity factor (f_a), usually employed in the characterization of coals [35–37] and heavy fractions [11,12], was calculated according to Poveda [34] (Table 14.2). These results show, as a natural consequence of the presence of asphaltenes, that the residues are more aromatic than the condensates. Moreover, both condensates and residue R were more aromatic than condensates and residues T and G. In the same way, R samples had a higher density than T and G, while the T condensates had a lower density and lower f_a.

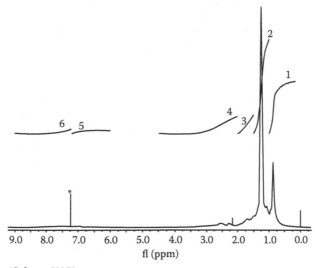

*Solvent CHCl$_3$
1. (0.1–1.0 ppm) paraffinic H γ and more to aromatic systems
2. (1.0–1.5 ppm) paraffinic H β to aromatic systems
3. (1.5–2.0 ppm) naphthenic H β to aromatic systems
4. (2.0–4.5 ppm) paraffinic and naphthenic H linked to aromatic systems
5. (6.0–7.2 ppm) aromatic H linked to monoaromatic rings
6. (7.2–9.0 ppm) aromatic H linked to di to polyaromatic rings

FIGURE 14.13 ^{1}H-NMR spectrum of C2G and assignments.

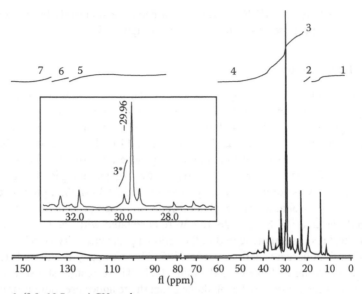

1. (3.0–18.5 ppm) CH_3 carbon atom type
2. (18.5–21.5 ppm) CH_3 alpha to aromatic rings
3. (21.5–50.0 ppm) CH_2 and CH napthenic and paraffinic
   3*. (30.30–29.75 ppm) CH_2 long chains
4. (50.0–60.0 ppm) CH tertiary and C quaternary paraffinic
5. (85.0–129.2 ppm) C benzo-naphthenic and aromatic carbon
   bridge of three aromatic rings
6. (129.2–137.0 ppm) C_{ar} head bridge between aromatic rings
7. (137.0–165.0 ppm) C alpha to S or N in benzo-structures

FIGURE 14.14 ^{13}C-NMR spectrum of C2G and the assignments.

As it was expected, AC and f_a have positive correlations as shown in Figure 14.15. This demonstrates that the AC parameter is an indicator of aromaticity. On the other hand, the ratio between 30.30–29.75 ppm (CH_2 in long aliphatic chains) and 18.5–21.5 ppm (CH_3 adjacent to aromatic systems) from ^{13}C-NMR showed a positive correlation with the ACL (Figure 14.16). The values for CH_2/CH_3 are reported in

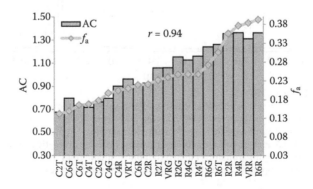

FIGURE 14.15   Correlation between aromaticity factor (by ^{13}C-NMR) and AC (by IR).

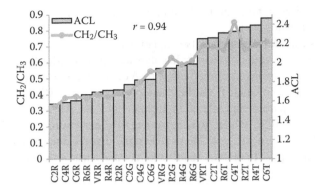

FIGURE 14.16   Correlation between the ratio CH_2/CH_3 from ^{13}C-NMR and ACL from IR.

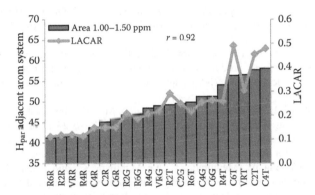

FIGURE 14.17   Correlation between H adjacent paraffinic protons to aromatic systems from ^1H-NMR and LACAR from IR.

Table 14.2. Finally, the LACAR parameter was correlated with the area 1.0–1.5 from ^1H-NMR (Figure 14.17). According to Figure 14.13, this region is assigned to paraffinic protons adjacent to aromatic systems. In general, T samples had higher values of this region than the G and R samples.

14.4   CONCLUSIONS

Eight VR from typical Colombian crude oils were fractioned into IBP-603°C, IBP-645°C, and IBP-687°C using short-path distillation. Fractions and VR were characterized with IR spectroscopy and analyzed by using HCA and PCA. Both HCA and PCA allowed distinguishing samples according to molecular features using two IR spectral regions, 2800–3000 cm⁻¹ and 700–900 cm⁻¹. According to PCA, samples were better clustered within the aromatic region 700–900 cm⁻¹ (five clusters) than in the aliphatic region 2800–3000 cm⁻¹ (three clusters).

The successful interpretation of mid-IR–ATR spectra through HCA and PCA allowed proposing four structural parameters: AC, Al, ACL, and LACAR. ACL explained the discrepancies in the distillation yields. A higher ACL is an indication of a higher distillation yield.

From the structural parameters, two equations were proposed to calculate density and MCR. From the equation to calculate density, one can infer that the parameters Al and ACL have positive correlations with API, whereas AC has a negative correlation. On the other hand, from the equation to calculate MCR, it is possible to conclude that the parameters related to aromaticity, LACAR and AC have positive correlations with MCR, while ACL has a negative correlation. Finally, according to IR and NMR data, it was possible to determine that residues R and G had higher aliphatic chain lengths than their condensate counterparts, whereas for T that trend was not observed.

REFERENCES

1. Martínez, A. 2011. *Ecopetrol, Energía Limpia para el Futuro: Macroeconomía y petróleo en Colombia*, Commemorating Book of 60 years of Ecopetrol. Bogota: Villegas Editores.
2. Molina, D., Navarro, U., and Murgich, J. 2007. Partial least-squares (PLS) correlation between refined product yields and physicochemical properties with the 1H nuclear magnetic resonance (NMR) spectra of Colombian crude oils. *Energy Fuels* 21: 1674–1680.
3. Sami-Nashawi, I., Malallah, A., and Al-Bisharah, M. 2010. Forecasting world crude oil production using multicyclic Hubbert model. *Energy Fuels* 24:1788–1800.
4. Hsu, C.S., McLean, M.A., Qian, K. et al. 1991. Online liquid-chromatography mass-spectrometry for heavy hydrocarbon characterization. *Energy Fuels* 5:395–398.
5. Jarne, C., Cebolla, V., Membrado, L., Le Mapihan, K., and Giusti, P. 2011. High-performance thin-layer chromatography using automated multiple development for the separation of heavy petroleum products according to their number of aromatic rings. *Energy Fuels* 25:4586–4594.
6. McKenna, A., Purcell, J., Rodgers, R., and Marshall, A. 2010. Heavy petroleum composition. 1. Exhaustive compositional analysis of Athabasca bitumen HVGO distillates by Fourier transform ion cyclotron resonance mass spectrometry: A definitive test of the Boduszynski model. *Energy Fuels* 24:2929–2938.
7. Cho, Y., Witt, M., Hwan Kim, Y., and Kim, S. 2012. Characterization of crude oils at the molecular level by use of laser desorption ionization Fourier-transform ion cyclotron resonance mass spectrometry. *Anal. Chem.* 84:8587–8594.
8. Owens, P., and Ryder, A. 2011. Low temperature fluorescence studies of crude petroleum oils. *Energy Fuels* 25:5022–5032.
9. Muller, H., Pauchard, V., and Hajji, A. 2009. Role of naphthenic acids in emulsion tightness for a low total acid number (TAN)/high asphaltenes oil: Characterization of the interfacial chemistry. *Energy Fuels* 23:1280–1288.
10. Avid, B., Sato, S., Takanohashi, T., and Saito, I. 2004. Characterization of asphaltenes from Brazilian vacuum residue using heptane–toluene mixtures. *Energy Fuels* 18:1792–1797.
11. Pereira de Oliveira, L., Trujillo-Vazquez, A., Verstraete, J.J., and Kolb, M. 2013. Molecular reconstruction of petroleum fractions: Application to vacuum residues from different origins. *Energy Fuels* 27:3622–3641.
12. Morgan, T., Alvarez-Rodriguez, P., George, A., Herod, A., and Kandiyoti, R. 2010. Characterization of Maya crude oil maltenes and asphaltenes in terms of structural parameters calculated from nuclear magnetic resonance (NMR) spectroscopy and laser desorption–mass spectroscopy (LD–MS). *Energy Fuels* 24:3977–3989.

13. Gaspar, A., Zellermann, E., Lababidi, S., Reece, J., and Schrader, W. 2012. Characterization of saturates, aromatics, resins, and asphaltenes heavy crude oil fractions by atmospheric pressure laser ionization Fourier transform ion cyclotron resonance mass spectrometry. *Energy Fuels* 26:3481–3487.
14. González, S.F., Carrillo, J., Núñez, M., Hoyos, L.J., and Giraldo, S. 2010. Modified design for vacuum residue processing. *Cienc. Tecnol. Futuro* 4:57–69.
15. Carrillo, J.A., and Corredor, L.M. 2008. Deep thermal conversion of a demetalized oil: Study of the deasphalting process using *n*-hexane. *Ing. Quim.* 40:92–104.
16. Speight, J.G., and Özüm, B. 2002. *Petroleum Refining Processes.* New York: Dekker.
17. Parra, M.J., León, A.Y., and Hoyos, L.J. 2010. Separation of fractions from vacuum residue by supercritical extraction. *Cienc. Tecnol. Futuro* 4:83–90.
18. Chen, S.L., Jia, S.S., Luo, Y.H., and Zhao, S.Q. 1994. Mild cracking solvent deasphalting: A new method for upgrading petroleum residue. *Fuel* 73:439–442.
19. Pang, W., Lee, J.K., Yoon, S.H. et al. 2010. Compositional analysis of deasphalted oils from Arabian crude and their hydrocracked products. *Fuel Process. Technol.* 91:1517–1524.
20. León, A.Y., and Parra, M.J. 2010. Determination of molecular weight of vacuum residue and their SARA fractions. *Cienc. Tecnol. Futuro* 4:101–112.
21. Zuñiga-Liñan, L., Nascimento-Lima, N.M., Wolf-Maciel, M.R. et al. 2011. Correlation for predicting the molecular weight of Brazilian petroleum residues and cuts: An application for the simulation of a molecular distillation process. *J. Pet. Sci. Eng.* 78: 78–85.
22. Azcan, N., and Yilmaz, O. 2013. Microwave assisted transesterification of waste frying oil and concentrate methyl ester content of biodiesel by molecular distillation. *Fuel* 104:614–619.
23. Sbaite, P., Batistella, C.B., Winter, A. et al. 2006. True boiling point extended curve of vacuum residue through molecular distillation. *Pet. Sci. Technol.* 24:265–274.
24. Zuñiga-Liñan, L., Savioli-Lopes, M., Wolf-Maciel, M.R. et al. 2010. Molecular distillation of petroleum residues and physical–chemical characterization of distillate cuts obtained in the process. *J. Chem. Eng. Data* 55:3068–3076.
25. Boduszynski, M.M. 1987. Composition of heavy petroleums. 1. Molecular weight, hydrogen deficiency, and heteroatom concentration as a function of atmospheric equivalent boiling point up to 1400°F (760°C). *Energy Fuels* 1:2–11.
26. Sbaite, P., Batistella, C.B., Winter, A. et al. 2006. True boiling point extended curve of vacuum residue through molecular distillation. *Pet. Sci. Technol.* 24:265–274.
27. Maciel Filho, R., Batistella, C.B., Sbaite, P. et al. 2006. Evaluation of atmospheric and vacuum residues using molecular distillation and optimization. *Pet. Sci. Technol.* 24:275–283.
28. Poveda, J.C. 2003. Caracterización Estructural de Fracciones Pesadas del Petróleo Mediante Técnicas Espectroscópicas. Tesis de Maestría en Química, Universidad Industrial de Santander, Bucaramanga.
29. Molina, D.R. 2008. Composición molecular promedio de crudos Colombianos y sus fondos de vacío y asfaltenos y su aplicación en el desarrollo de un modelo para optimizar una unidad de desasfaltado. Tesis de Doctorado en Química, Universidad Industrial de Santander, Bucaramanga.
30. Orrego-Ruiz, J.A., Guzmán, A., Molina, D., and Mejía-Ospino, E. 2011. Mid-infrared attenuated total reflectance (MIR–ATR) predictive models for asphaltene contents in vacuum residua: Asphaltene structure–functionality correlations based on partial least-squares regression (PLS-R). *Energy Fuels* 25:3678–3686.
31. Almeida, J.A.S., Barbosa, L.M.S., Pais, A.C., and Formosinho, S.J. 2007. Improving hierarchical cluster analysis: A new method outlier detection and automatic clustering. *Chemom. Intell. Lab. Syst.* 87:208–217.

32. Fernández-Varela, R., Suárez-Rodríguez, D., Gómez-Carracedo, M.P. et al. 2005. Screening the origin and weathering of oil slicks by attenuated total reflectance mid-IR spectrometry. *Talanta* 68:116–125.
33. Gómez-Carracedo, M.P., Fernández-Varela, R., Ballabio, D., and Andrade, J.M. 2012. Screening oil spills by mid-IR spectroscopy and supervised pattern recognition techniques. *Chemom. Intell. Lab. Syst.* 114:132–142.
34. Poveda, J.C., and Molina, D.R. 2012. Average molecular parameters of heavy crude oils and their fractions using NMR spectroscopy. *J. Pet. Sci. Eng.* 84–85:1–7.
35. Mejía-Ospino, E., and Orrego-Ruiz, J.A. 2012. *Estudio de la Estructura de Carbones Colombianos por FTIR: Análisis de Extractos Orgánicos.* Editorial Académica Española, Madrid, Spain.
36. Orrego, J.A., Cabanzo-Hernández, R., and Mejía-Ospino, E. 2010. Structural study of Colombian coal by fourier transform infrared spectroscopy coupled to attenuated total reflectance (FTIR–ATR). *Rev. Mex. Fis.* 56(3):251–254.
37. Orrego-Ruiz, J.A., Cabanzo, R., and Mejía-Ospino, E. 2011. Study of Colombian coals using photoacoustic Fourier transform infrared spectroscopy. *Int. J. Coal Geol.* 85:307–310.

Index

Page numbers followed by f and t indicate figures and tables, respectively.